Innovations of Green Process Engineering for Sustainable Energy and Environment 2019

Topical Conference at the 2019 AIChE Annual Meeting

Orlando, Florida, USA
10-15 November 2019

ISBN: 978-1-7138-0564-9

Printed from e-media with permission by:

Curran Associates, Inc.
57 Morehouse Lane
Red Hook, NY 12571

Some format issues inherent in the e-media version may also appear in this print version.

Printed with permission by Curran Associates, Inc. (2020)

AIChE
120 Wall Street, FL 23
New York, NY 10005-4020

Phone: (800) 242-4363
Fax: (203) 775-5177

www.aiche.org

Additional copies of this publication are available from:

Curran Associates, Inc.
57 Morehouse Lane
Red Hook, NY 12571 USA
Phone: 845-758-0400
Fax: 845-758-2633
Email: curran@proceedings.com
Web: www.proceedings.com

TABLE OF CONTENTS

573934 Strategies for Assessing and Improving Microbial Utilization of Lignin-Derived Monomers

The economic viability of the biorefinery concept is limited by the valorization of lignin. One possible method of lignin valorization is biological upgrading with aromatic-catabolic microbes. In conjunction, lignin monomers can be produced by fast pyrolysis and fractionation. However, biological upgrading of these lignin monomers is limited by low water-solubility. Here, we address the problem of low water-solubility with an emulsifier blend containing approximately 70 wt% Tween® 20 and 30 wt% Span® 80. *Pseudomonas putida* KT2440 grew to an optical density (OD_{600}) of 1.0±0.2 when supplied with 1.6 wt% emulsified phenolic monomer-rich product produced by fast pyrolysis of red oak. The emulsifier dose was 0.076±0.002 g emulsifier blend per g of phenolic monomer-rich product. This approach partially mitigated the toxicity of the water-soluble model phenolic monomer *p*-coumarate to the microbe. This study provides a proof of concept that processing of biomass-derived phenolics to increase aqueous availability can enhance microbial utilization.

576437 Developing Bacterial Pelletized Culture to Enhance Yield and Simplify Harvest for Lignin Bioconversion

Pelletized culture has been long sought after for filamentous fungi fermentation due to its several advantages such as easier separation, decreased viscosity. However, pelletized culture was achieved little in bacteria, which prevents the further increase of bioproduct yield and the utilization of lower cost filtration harvest. We have developed a novel pelletized culture system for Gram positive bacteria Rhodococcus opacus PD630 and applied the new system to lignin bioconversion. Moreover, we have identified several factors impacting the pelletization performance and thus optimized the system to increase lignin to lipid conversion. We found that inoculum size, centrifugation speed, soluble solid content (SSC), glucose content, and nitrogen content each had an impact on the pellet formation. The results showed that pellet percentage increased with the increase of SSC to 41.91%, when 50g/L SSC was added into medium. And the pellet percentage varied from 29.31% to 50.63%, when glucose was added with a concentration increase from 0g/L to 8g/L. The highest percentage of pelletization was achieved at 2g/L glucose. Further analysis indicated that the lipid content within pellets remains stable even SSC increased up to 50g/L, whereas the lipid content of the cells in solution phase varies depending on the SSC. Overall, this is the first report of bacterial pelletized culture without addition of flocculant. When further optimized, the innovative platform provides a new solution for improving bioproduct yield and reducing harvest cost in bacterial cultures. In addition, as applied to lignin bioconversion, the technology has potential to transform bioproduction from lignin by simplifying harvest and increasing yield.

574100 Unlocking the Genetic Potential of *Rhodopseudomonas Palustris* CGA009 for Lignin Valorization

Non-model microorganisms have unique and robust biochemical processes that are unavailable in their model counterparts. Harnessing their unique capabilities have the potential to enable the sustainable synthesis of numerous products; yet, the ability to precisely and reliably engineer these microbes is hampered by their innate repair and immune systems as well as a lack of synthetic biology tools. The purple non-sulfur bacterium *Rhodopseudomonas palustris* CGA009 employs four different metabolic modes, photoautotrophic, photoheterotrophic, chemoautotrophic, and chemoheterotrophic, to adapt to a diversity of environmental conditions. In addition to fixing carbon dioxide during autotrophic growth, the bacterium can also degrade recalcitrant aromatic compounds derived from lignin, using both aerobic and anaerobic pathways. *R. palustris* is the only bacterium that encodes two carbon dioxide-fixing RubisCO-like proteins and it is one of two known bacteria that encodes all three nitrogen-fixing nitrogenase isozymes. Purple non-sulfur bacteria perform anoxygenic photosynthesis which allows oxygen-sensitive nitrogen fixation to occur at the same time. Furthermore, even in very low intensity light *R. palustris* can remain metabolically active in a non-growing state for months. Published engineering efforts that unlock *R. palustris*' genetic potential to valorize lignin breakdown products have been minimal, possibly due to the bacterium's intrinsic antibiotic resistance, the genetic instability of exogenous DNA, and the lack of precise orthogonal regulators of gene expression. Multiple tracks are being pursued to address these limitations. In addition to characterizing the microbe's antibiotic resistance, genes for antibiotic-modifying enzymes have been removed from the genome. Methylation patterns of heterologous plasmids are being explored, already with a dramatic improvement in transformation efficiency. Neutral sites for genomic integration of synthetic devices are being characterized. The impact of genes in the microbe's mutagenic translation synthesis system on the genetic stability of exogenous DNA are also being investigated through knockout strains. Finally, synthetic biology tools for transcriptional regulation are being developed to explore *R. palustris*' catabolic pathways for lignin-derived compounds and bioplastic production. This work will contribute to the development of this promising bacterium as a biotechnology platform and to the generation of important design guidelines for engineering non-model microorganisms.

567727 Improving Resistance of Hyperthermophilic Laccase to Ionic Liquids for *in Situ* Lignin Valorization

Lignin makes up 20-30% of plant biomass and is the most abundant aromatic polymer found in nature. Converting lignin from a waste product to value-added chemicals and precursors can increase the revenue stream of cellulosic biorefineries. Ionic liquids (ILs) have been subjects of numerous studies in recent years for their ability to fractionate biomass during pretreatment. Some ILs that are capable of fractionating biomass are also biocompatible with enzyme activity. Given the unique properties of these ILs, there is a unique opportunity to develop new strategies for lignin valorization *via* biocatalysis in ILs. Thermophilic enzymes have demonstrated improved resistance to IL inhibition when compared to their mesophilic counterparts. To that end, we screened the biocompatibility of 3 ILs with a hyperthermophilic laccase from *Thermus thermophilus*. Despite the extreme thermophilicity of this laccase (T_{opt} = 92 °C), it demonstrated sensitivity to low concentrations of cholinium lysinate ([Ch][Lys]) and diethylammonium hydrogensulfate ([DEA][HSO_4]). The IL, 1-ethyl-3-methylimidazolium acetate, [C2C1Im][OAc] was observed to be more biocompatible, with severe inaction only observed in 10% [C2C1Im][OAc]. Docking simulations show that these ILs favor binding to a location near the active site that disrupts substrate binding, leading to the perceived sensitivity to the ILs. Using a rational design approach led to the improved activity of this laccase in aqueous ILs, foreshadowing its capabilities to valorize lignin in ILs after biomass pretreatment. Collectively, these results demonstrate the analysis of laccase-IL interactions and the improvement of laccase activity in ILs *via* a rational design approach.

574110 Computational Pathway Design for Funneling Lignin Intermediates to Aromatic Products

The heterogeneity of the aromatic products from lignin catalytic depolymerization is one of the major challenges for lignin valorization. Microbes have evolved many catabolic pathways to tackle the challenge by funneling such heterogeneous intermediates to a few central aromatic products, which lead further to intra- or extradiol ring opening to produce value-added chemicals. However, such funneling pathways are only well-characterized in a few organisms such as *Sphingobium* sp. SYK-6 and *Pseudomonas putida* KT2440. Further effort to explore novel bacterial funneling pathways is necessary. Herein, we apply the *de novo* pathway design tool (novoStoic [1]) to computationally prospect all possible pathways for lignin-derived mono- and biaryls. First, we augment the MetRxn database with a new dataset of elementally balanced reaction operators using the automated atom mapping based reaction rule extraction procedure termed rePrime. For each reaction, the rePrime procedure identifies and captures as reaction rules the molecular graph topological changes underpinning the substrate to product graph conversion. A reaction rule captures the location of active reaction centers affected by the conversion of substrates to products. The reaction rules and known reactions are then operated upon a mixed integer linear programming algorithm (novoStoic) to identify a mass-balanced biochemical network that converts a source to a target metabolite while minimizing the number of *de novo* steps. We demonstrate the application of novoStoic in a few case studies such as designing shorter pathways from ferulate to protocatechuate and exploring C-C cleavage pathway of guaiacylglycerol--guaiacyl ether. By exploring the uncharted chemical space enabled by enzyme promiscuity, novoStoic paves the way to identify more carbon/energy efficient lignin funneling pathways with minimal heterologous enzymes to engineer, while exploring the organism's potential underground metabolism.

References

[1] Kumar, A., et al. Pathway design using de novo steps through uncharted biochemical spaces. *Nature communications* 9.1 (2018): 184.

575341 Enhancement of Polyhydroxyalkanoate Production By Co-Feeding Lignin Derivatives with Glycerol in *Pseudomonas Putida* KT2440

Pseudomonas putida has great potential for biosynthesis of polyhydroxyalkanoate (PHA) to produce biodegradable plastics. In this study, with co-feeding lignin derivatives (i.e. benzoate, vanillin, and vanillic acid) in addition to glycerol, the cell dry weight and PHA content at 72 h improved by 14%-20% and 49.1%-72.6%, respectively. GC-MS analysis of PHA compositions revealed that the addition of lignin derivatives to glycerol decreased the distribution of long chain monomers (e.g. C10 and C12) of PHA while increased the short chain monomers (e.g. C6 and C8). A newly developed ^{1}H / ^{13}C HMBC method that directly examined cell samples without the PHA extraction process confirmed these findings. Further optimization of benzoate/glycerol/nitrogen ratio revealed that the addition of benzoate increased intracellular NADH and NADPH concentrations, indicating the strengthened oxidative stress, consistent with the improvement of both PHA and cell biomass yields. In addition, ^{1}H, ^{13}C and ^{31}P NMR metabolite analysis suggested that the co-feeding of benzoate and glycerol enhanced glycerol consumption, and potentially improved the ED pathway. Overall, this study provides an effective co-carbon feeding strategy to improve PHA yield and lignin conversion.

574051 Conversion of Lignin into Biodegradable Plastic By *Rhodopseudomonas Palustris*

A sustainable alternative to conventional plastics is polyhydroxyalkonoates (PHAs), biopolymers produced by microorganisms as energy storage granules in certain environmental conditions. PHAs have similar properties as synthetic plastics, are the only 100% truly biodegradable polymer, and can be produced from a variety of substrates. As the most abundant storage compound in bacteria, poly-3-hydroxybutanoate (PHB) is a widely studied short-chain-length PHA. However, PHAs are inhibited by higher production costs due primarily to the selected strain, carbon source, and the need for more efficient extraction methods. Thus, the goal of this study was to engineer the metabolically versatile *Rhodopseudomonas palustris* CGA009 to produce PHB from lignocellulosic biomass as a renewable carbon source and promote the valorization of PHAs. As part of this study, *R. palustris*' native metabolic and regulatory PHB production pathways on lignin breakdown products (LBPs) were evaluated. Results confirmed that *R. palustris* grows both anaerobically and aerobically on the major LBPs p-coumarate and coniferyl alcohol, as well as on Kraft lignin supplemented with acetate. Furthermore, chemical oxygen demand testing confirmed *R. palustris* catabolizes these carbon sources. Using aqueous two-phase extraction and gas chromatography-mass spectrometry, we found that *R. palustris* does generate PHB without nutrient deprivation, but that nitrogen starvation results in significantly higher accumulations. Preliminary results show that *R. palustris* produces approximately 5g PHB in 13.5mL 1mM LBP after four days of nitrogen starvation, while the optimum timing for maximum production is being investigated. *R. palustris*' PHB gene expression was analyzed via qPCR to further characterize its native pathways. Our ongoing genetic engineering efforts utilize these findings to improve PHB production in *R. palustris* by creating new strains with optimized granule associated proteins. Additionally, our continuing efforts to develop an efficient extraction method specific for this strain using non-chlorinated and cost-effective solvents will further enhance PHB yields. Ultimately, this study is novel in that it combines genetic engineering and efficient fermentation process design to advance PHB production from a non-model and metabolically flexible bacteria using lignin as a renewable carbon substrate through a multifaceted approach: (i) characterizing *R. palustris*' native growth and PHB production on LBPs, (ii) genetically engineering *R. palustris*' granule associated proteins for the overproduction of PHB, and (iii) designing an effective PHB recovery method specific for this strain. These efforts thwart the two major production costs for PHB, expensive carbon source and extraction techniques, that are preventing PHB production from becoming cost-effective replacements for some petroleum-based plastics.

577450 Elucidating Lignin Structure and Relationships from Naturally Variant *Populus Trichocarpa*

Lignocellulosic biomass is a promising renewable alternative to replace fossil fuels. The primary challenge to utilizing lignocellulosic material is its inherent recalcitrance to enzymatic hydrolysis. This recalcitrance is typically overcome by thermal and/or chemical pretreatment. Pretreatment performance is greatly influenced by lignin, a complex biopolymer. Therefore, it is important to understand the structure of native lignin. Due to its completely sequenced genome, it is possible to elucidate the genetic basis of lignin variation in *Populus trichocarpa* using phenotype-to-genotype correlation. To this effect, we have analyzed over 200 unique genotypes of *Populus trichocarpa* for lignin phenotype by HSQC NMR. This analysis has provided insights to the relative abundance and variation of syringyl, guaiacyl, p-hydroxyphenyl, and p-hydroxybenzoate units, as well as aryl ether, phenylcoumaran, and resinol linkages. Several correlations were also observed, such as syringyl:guaiacyl ratio to p-hydroxybenzoate units, aryl ether, and phenylcoumaran linkages. In addition, a gene expression analysis has identified 22 genes which are co-expressed with transcription factors known to impact cell wall biosynthesis. A variety of phenotypes, including phenotype extremes, are being further investigated for additional lignin properties such as molecular weight and antioxidant activity. Preliminary results suggest there may be a relationship between molecular weight and aryl ether linkages, as well as antioxidant activity and resinol linkages.

576048 Separation of Lignin Streams after Alkali and Ionic Liquid Pretreatment Using Tangential Flow Filtration

Lignin is Nature's most abundant aromatic polymer, and replacement of petroleum-based chemicals with those derived from lignin represents a virtuous challenge for chemists. After biomass pretreatment under basic condition, most or significant amount of the lignin will be stripped from the feedstock and retained in the aqueous stream. In this work, lignin-rich stream was obtained from alkali and ionic liquid (IL) pretreatment process on corn stover. After pretreatment, the slurry was cooled down and solids removed. Three different technologies are being investigated for solids removal: ultrasonic separation (US), tangential flow filtration (TFF) and centrifugation. Ultrasonic separation is being used to remove fines (i.e., particles < 100 microns) for improved TFF membrane permeability or increase flow rates for centrifugation. Ultrasonic separation of fines can be achieved with low energy due to the high acoustic contrast between the lignocellulosic solids and water medium. TFF was also employed for the fractionation of lignin molecules. Samples were taken from processing streams and analyzed using different techniques. In general, weight-average and polydispersity decreased with the decrease of membrane pore size. Lignin properties from the two different pretreatment processes were compared to understand the pretreatment mechanisms with distinct catalysts.

#567406

REPLACING COAL USED IN STEELMAKING WITH BIOCARBON FROM HYDROLYSIS LIGNIN -TECHNO-ECONOMIC EVALUATION

Juha Hakala [1], Petteri Kangas [1], Martin Björnström [2], Timo Fabritius [3] and Pertti Koukkari [1]
[1] VTT Technical Research Centre of Finland, Espoo, Finland
[2] Helsinki
[3] University of Oulu

Abstract

Iron and steel industry is the second largest industrial emitter in the world after the cement industry. In 2017 Global crude steel production direct CO_2 emissions of 2 billion tonnes represented 23.5% of industrial emissions in the world. Bio-reducing agents provide the possibility of achieving a rapid reduction in fossil CO_2 emissions in the steel industry. Forest-sector residues can be converted by slow pyrolysis to biocarbon for reduction purposes, which can be an asset for both the forest and metal industry. This study focused on the evaluation of the production of biocarbon from hydrolysis lignin, a side-product from 2nd generation bioethanol process, a) produced in a stand-alone plant close to a steel mill and b) produced in a pulp and paper mill integrated plant. The conceptual level techno-economic analysis indicates that biocarbon made of hydrolysis lignin at an integrated pulp and paper mill site will provide the potential option for bio-based carbon production within the forest to metal value chain, providing a substantial source, at an annual potential of 44 k tonnes of biocarbon. Hydrolysis lignin could thus provide a significant biocarbon source for the pulverized coal injection (PCI) in the blast furnace, in addition to other bio-based side-streams (e.g. bark, wood/forest chips and sawdust). The economic feasibility will depend on emission trading, fossil coal price and on integration of the excess energy from pyrolysis and possibly available low-grade heat at the biocarbon production site.

1 Introduction

Most of CO_2 emissions in steel production occur during the reduction of iron ore to hot metal through the use of coal and coke. Renewable carbon containing material from the forest sector is seen as an asset for reducing fossil emissions of steel industry. This bio-based carbon can be used as bio-reducing agents in the short term for providing rapid reduction of CO_2 emissions. New process concepts based on usage of renewables for energy and reduction can be implemented in the long run (Suopajärvi et al., 2018; Ng et al., 2018), such as hydrogen reduction and/or the use of electricity in plasma technology. However, raw biomass as such is unsuitable for applications in iron and steel making and thermochemical conversion into biocarbon is required. Slow pyrolysis is the most promising technology to upgrade biomass as a metallurgical reducing agent as indicated in a recent review (Suopajärvi et al., 2018).

The key focus of current study is to evaluate direct substitution of fossil-based pulverized coal injection (PCI) in the blast furnace with biocarbon from sustainable sources. One of tempting raw material is hydrolysis lignin, which is a side-product from 2nd generation bioethanol process.

An example of such the bioethanol plant is St1 Biofuels Cellunolix® bioethanol plant in eastern Finland (Yamamoto, 2018). The amount of bioethanol production utilising lignocellulosic side streams (such as sawdust, bagasse and corn stover) is increasing in the future. Accordingly, the amount of hydrolysis lignin available will increase. In addition, this study focused on creating an industrial symbiosis between bio-based / forest industry and metals processing industry.

2 Methods and materials

2.1 Methodology, scenarios and assumptions

A conceptual level techno-economic assessment was conducted for evaluating the feasibility of biocarbon production from lignin. Three optional scenarios were evaluated. Thermochemical conversion process was fixed as slow pyrolysis. Biocarbon is assumed to replace part of PCI coal injected into the blast furnace. Mass and energy balances were calculated using process simulation software, Balas (VTT, 2013) for evaluating the slow pyrolysis process and Wingems (Valmet, 2016) for conducting process integration studies for a kraft pulp mill. Production costs were evaluated during the economic assessment in a Finnish context (location, prices, etc.). Key process parameters were obtained from the literature. The experimental proof and arrangements for bio-reducer production from hydrolysis lignin were carried out concurrently and are presented in detail by Toloue Farrokh et al. (2019).

Stand-alone production refers to the scenarios (L50, L90) where the biocarbon pyrolysis unit is located on the site of a steel/iron plant on the north-west coast of Finland. Lignin is transported by rail to the production site from the location within 200 km. In L50 scenario lignin is received as moist (dry content 50 wt%) and in L90 scenario as dried (90 wt%). The biocarbon manufacturing process is not heat integrated with the steel plant itself. The hot pyrolysis gases are burned in order to provide steam, used in the first hand for lignin drying purposes. The effect of remaining surplus steam is evaluated by setting the electricity generation yield to 35% of the available heat load. The power plant itself is not included in the scenario estimate.

Pulp and paper mill integrated biocarbon production refers to the scenario (L50-I) where the production of biocarbon takes place at the pulp and paper mill, utilizing integration possibilities. Moist lignin (50 wt%) is assumed to be available on a site at the pulp and paper mill (modern Nordic Kraft mill located in west coast of Finland). Drying of lignin prior pyrolysis utilizes hot water and steam, each meeting 50% of the necessary heat load. It is assumed that the pulp mill can provide the hot water and steam. The hot pyrolysis gases (net) are redirected for further use as a fuel for the lime kiln at the pulp mill, replacing original fuel. To cover endothermic reactions during pyrolysis, some of the generated pyrolysis gases are burned to provide the necessary heat load. From these flue gases some steam is generated for drying, reducing the need for steam from the pulp mill. No excess steam is available. The biocarbon product is then transported by rail to the iron/steel plant within 160 km.

For all the scenarios mentioned hydrolysis lignin energetic valuation is based on the 20 €/MWh (Statistics Finland, 2018) and the applied electricity price of €35/MWh is based on the principle that, in Finland, power plant owners may purchase electricity at prime cost rates. The biocarbon production volume is 44 200 tonnes per year, which would cover ca. 10 wt% of PCI coal injected to blast furnace (in Raahe SSAB steelworks). All the assumptions, detailed information of the pulp and paper mill reference and related integration parameters, as well as techno-economic calculations can be found from Hakala et al. (2019).

2.2 *Raw material*

Hydrolysis lignin, the side product of the fermentation process producing bioethanol from sawdust, was chosen as the raw material, as further utilization of lignin is an important target for the ethanol producers (Yamamoto, 2018). Currently, the hydrolysis lignin can be utilized to produce heat or steam at the power plant. In the hydrolysis lignin, the actual lignin content varies 70-80 wt% and residual cellulose 20-30 wt% of the dry matter, while the ash and sulphur contents are low. Some amounts of sugar, acids and degradation products are also present. The lignin has low reactivity (Yamamoto 2018). In the scenarios chosen in this study we use an annual lignin production level of ~100 000 dry tonnes, assuming that a larger plant in Finland is built in the near future.

Hydrolysis lignin samples were provided by St1 for the project for analysis and pyrolysis experiments, carried out at the University of Oulu. Detailed analyses of the lignin raw material are presented in Toloue Farrokh et al. (2019). The lower heating value (LHV) on a dry basis was analysed as 23.9 MJ/kg and the higher heating value (HHV) as 25.2 MJ/kg, respectively. The ash content (dry basis) was 0.3 wt%. Elemental analysis of lignin samples in moles, based on Toloue Farrokh et al., on a dry basis, without ash, and scaled to 1,000 g is C 51.18, H 61.98, N 0.50, O 19.68 and S 0.03. This corresponds to weight percentages as C 61.47%, H 6.25%, N 0.70%, O 31.49% and S 0.09%.

3 Results

3.1 *Technical evaluation*

Material balances for the scenarios are presented in *Table 1*. The significant difference is the lignin arriving as dry in the scenario L90. Differences in pyrolysis and flue gases amounts are explained by the fact that in stand-alone scenarios L50 and L90 all pyrolysis gases are burned within the system limits in order to produce electricity, while in the integrated scenario major proportion of the hot pyrolysis gases are directed for further use in the lime kiln. Only a small proportion of pyrolysis gases are burned to cover the endothermic reactions in the pyrolysis itself.

Energy balances as lower heating values (LHV) used in production costs estimates calculations are presented in *Table 2*. Presented values are based on balances applying higher heating values (HHV), used by Balas simulation software. The thermal load received by the system in drying was evaluated, instead of the real drying energy required. The applied external energy in drying and the related losses are outside the system boundaries. From the balances reveal that lignin arriving as dry (90 wt%) in scenario L90 require no drying energy prior to the pyrolysis (the minimum dry content requirement for the raw material set as 90 wt%). Stand-alone scenario L50 requires drying, carried out with the internally recovered heat, showing higher losses than stand-alone scenario L90. The various energy sources for drying and preheating are listed separately at the bottom of the *Table 2*, covering also the losses. In the integrated scenario L50-I, 50% of the energy required is covered by hot water from the pulp and paper mill. The remainder is covered by steam, primarily using internally generated steam, with the remaining steam sourced from the pulp and paper mill.

Table 1. Material balances resulting from the modelling.

	Lignin L50 [k tonne/a]	Lignin L90 [k tonne/a]	Lignin L50-I [k tonne/a]
Feed:			
Raw material, as received	196	109	196
Air for burning	328	328	60
Sum (feed):	**525**	**438**	**257**
Outputs:			
Biocarbon (at 50 °C)	44.2	44.2	44.2
Pyrolysis gases	*burned*	*burned*	53.0
Flue gases	393	393	72.1
Vapour from drying	87.3	0	87.3
Sum (outputs):	**525**	**438**	**257**

Table 2. Simplified energy balances (LHV) resulting from the modelling.

	Lignin L50 [MW]	Lignin L90 [MW]	Lignin L50-I [MW]
Feed:			
Raw material, as received	73.1	80.5	73.1
Thermal load received by the system (drying)	*internal*	*internal*	7.4
Sum (feed):	**73.1**	**80.5**	**69.8**
Outputs:			
Biocarbon (at 50 °C)	49.7	49.7	49.7
Pyrolysis gases	N/A	N/A	28.0
System excess energy used for drying (steam)	*internal*	*internal*	1.10
System excess energy for other purposes (steam)	13.3	27.1	-
Losses	10.1	3.76	1.74
Sum (outputs):	**73.1**	**80.5**	**69.8**
Drying/preheating energy division, losses included: [*)]			
External energy (hot water)	-	-	7.23
External energy (steam)	-	-	6.12
Internal energy (steam)	14.5	0.74	1.10
SUM (drying/preheating energy):	**14.5**	**0.74**	**14.5**

[*)] *Energy distribution for drying purposes from different sources. Integrated scenario values are essential for the calculation of production costs scenarios.*

3.2 Pulp mill integration

The effect of biocarbon production on the energy balances of an integrated pulp and paper mill are: i) annual net electricity production of the integrate is reduced by 3 GWh, and ii) the amount of available heat for district heating is reduced by ~50 GWh per year. These reductions are due to the increased demand for process steam within the biocarbon production process.

These losses are taken into account and targeted to the biocarbon production costs in the scenario L50-I. The data from the pulp and paper mill considered is not based on exact operational data. A detailed evaluation of the results can be found from Hakala et al. (2019).

3.3 Economic feasibility

The production costs estimate for the stand-alone lignin scenarios L50 and L90 are €405/tonne and €379/tonne of biocarbon respectively, also taking into account the utilization of excess steam for electricity generation. A major proportion of the costs are raw material costs, lignin, valued according to the heating value as received.

The estimate for production cost for the integrated scenario L50-I is €275/tonne of biocarbon (*Figure 1*). The raw material cost is the predominant cost factor, the same cost value as in L50 scenario, due to the same dry content of received lignin. Savings in the costs are achieved by replacing part of the lime kiln's original fuel with hot pyrolysis gases. Capital expenditure is also a notable cost component, and is estimated to be a bit higher than for L50 (€53/tonne) and L90 (€49/tonne) scenarios. Production losses are caused by the reduction in net electricity output and in energy for district heating. Labour and logistic costs are clearly lower than for stand-alone scenarios, due to the synergies in labour in the integrated mill, the shorter transportation distance and transporting densified biocarbon instead of the moist raw material.

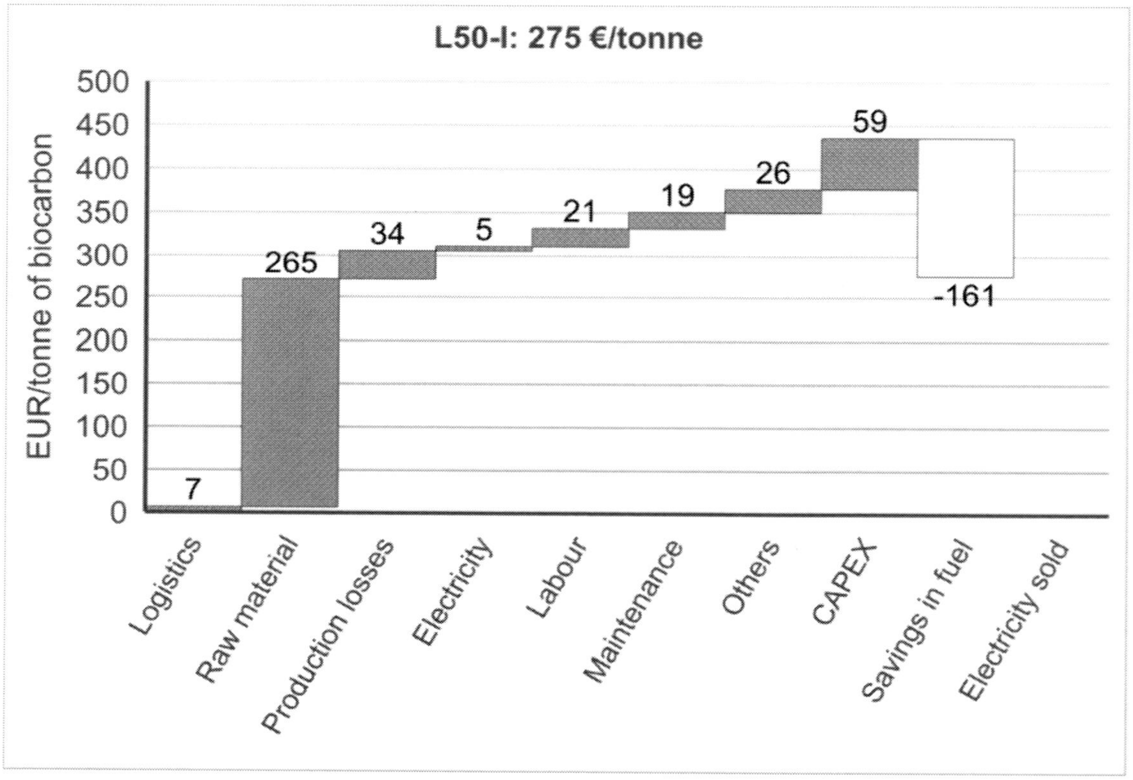

Figure 1. Production costs structure for the pulp mill integrated scenario.

3.4 Sensitivity analysis

A sensitivity analysis was based on stochastic simulation, using Monte Carlo Method. Simulation runs of 30,000 were conducted, in order to illustrate the effects of input parameter distribution on production costs of the integrated lignin scenario. The analysis tool @Risk was used (Palisade Corp., 2015). These input parameters were raw material price, cost of electricity, lignin moisture content, biocarbon yield, raw material feed as dry, transportation distance and total capital investment (TCI), parameter variations presented in *Table 3.* Raw material price is the most predominant factor, while TCI was the second.

Table 3. Selected and varied input parameters for the integrated lignin L50-I scenario's sensitivity analysis. Triangle distribution for input parameters was applied. The @Risk tool was used in the sensitivity analyses.

Parameter varied	MIN	STATIC	MAX
Lignin price [€/MWh]	15.0	20.0	25.0
Cost of electricity [€/MWh]	35.0	35.0	68.0
Lignin moisture content [wt%]	40.0%	50.0%	60.0%
Biocarbon yield [wt%] *)	41.8%	45.0%	48.2%
Raw material as dry [tonne/yr] **)	81,850	98,200	114,600
Transportation distance – biocarbon [km]	1	160	450
Total capital investment (TCI) [M€] ***)	19.2	27.4	35.7

*) *The biocarbon yield MIN value is set 93% and MAX to 107% of the static value*
**) *The lignin feed as dry is set to 83% and MAX to 117% of the static value yield*
***) *The TCI MIN value is set to 70% and MAX 130% of the static value*

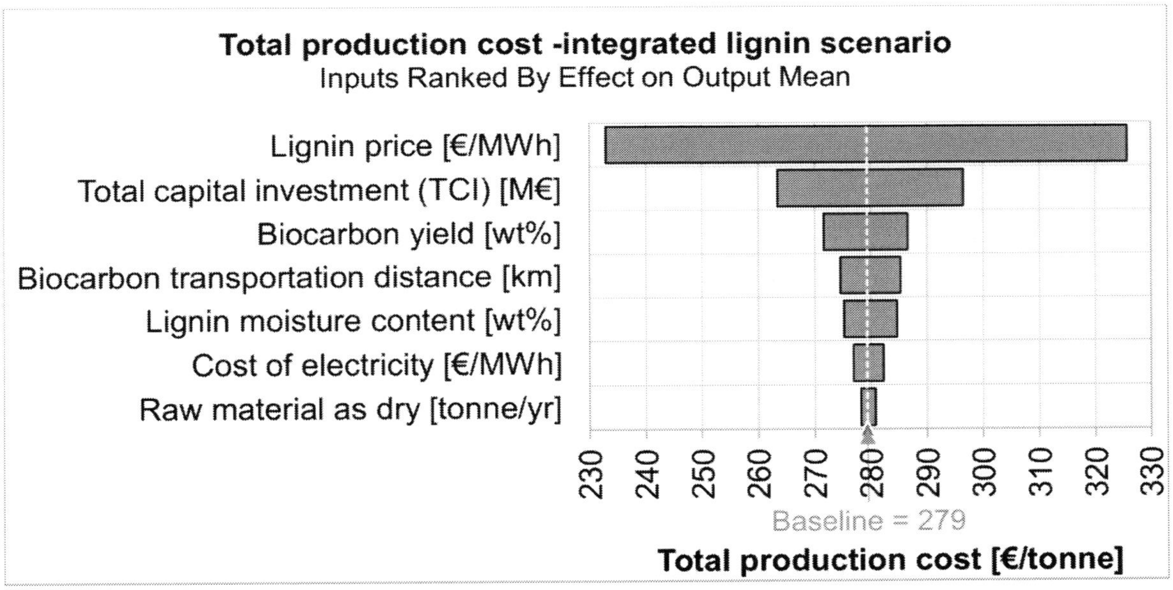

Figure 2. The sensitivity of biocarbon production costs to selected key cost factors. The baseline values are based on the given distributions and not on the static values.

4 Discussion

Hydrolysis lignin has been found to be applicable as a source of biocarbon production with a high enough carbon to oxygen ratio. The current acceptable level for a PCI coal replacement ratio with biocarbon is 10–20 wt%, while higher ratios will require further technological assessment of the detrimental elements affecting the blast furnace operation and solving certain technical limitations, such as sizing of PCI coal handling and pulverizing line, and re-designing the co-feed of coal and biocoal to the blast furnace. Physical and chemical properties must also be taken into account when designing the blast furnace biocarbon injection system (Hakala et al. 2019).

The cost of replacing coal and/or coke is a crucial factor for the success of new biocarbon-based process concepts. The estimated production costs for a stand-alone plant were ~ €380-405/tonne of biocarbon and the respective costs for an integrated process were ~€275/tonne of biocarbon. Previous studies indicate that the cost of bio-based reducers varies significantly based on the applied technology and quality of the bio-reducer (Suopajärvi 2018; 2014). The lowest estimate for production costs is ~ €110–180/tonne for torrefied wood. When biocarbon is considered, as in this study, the estimation is between ~ €220–510/tonne. The production costs for the integrated process in particular is close to the lower limit of these previous studies. The stand-alone production costs of biocarbon from lignin may be reduced, too, by heat and/or fuel gas integration with the steel mill. Production costs of all the scenarios may also be reduced by increasing the capacities.

Economic viability of the biocarbon utilization in reduction processes in metal industry depends of i) effective integration of the pyrolysis process and of the pyrolysis gases further utilization, ii) the price of fossil coal to be replaced, iii) as well as of the CO_2 emission trading scheme. EU emission allowance (EUA) pricing in Europe has been heavily increasing from mid 2017 (€5/tonne CO_2) to the present. The price level of €30/tonne CO_2 was almost met in July 2019 (Sandbag, 2019).

5 Conclusions

Based on the estimated biocarbon production costs, converted from hydrolysis lignin, the pulp and paper mill integrated scenario is approaching to be economically viable, providing costs reductions of over €100/tonne of biocarbon when compared to stand-alone process scenarios to be located close to the steel mill. There are significant benefits for process integration to a pulp (and paper) mill. These are the possibility to burn hot pyrolysis gases e.g. in the lime kiln, replacing original fuels, and utilization of excess low temperature heat such as hot water and steam for raw material drying. Hydrolysis lignin can also provide a substantial source of biocarbon, at an annual potential of 44,000 tonnes in the case presented.

Acknowledgements

This work was supported by the Business Finland and conducted as a part of the FOR&MET R&D project Added Value from Forest Industry for Metals Producing and Processing Integrates (2016–2019), as well as a part of the SYMMET R&D project Symbiosis of metals production and nature (2018–2020).

References

1. Hakala, J., Kangas, P., Penttilä, K., Alarotu, M., Björnström, M., & Koukkari, P. (2019) "Replacing Coal used in Steelmaking with Biocarbon from Forest Industry Side Streams". VTT Technology No. 351. VTT Technical Research Centre of Finland. Available at: https://doi.org/10.32040/2242-122X.2019.T351

2. Ng, K. W., Giroux, L. and Todoschuk, T. (2018) "Value-in-use of biocarbon fuel for direct injection in blast furnace ironmaking", Ironmaking and Steelmaking. doi: 10.1080/03019233.2018.1457837.

3. Palisade Corp. (2015) "@Risk". Risk Analysis Add-in for Microsoft Excel. Version 7.0.1., Professional Edition Build 396.

4. Sandbag (2019) "Carbon price viewer". Raw data from ICE via Quandl. Entered September 10th 2019. Available at: https://sandbag.org.uk/carbon-price-viewer/

5. Statistics Finland (2018) "001 Kotimaisten polttoaineiden käyttäjähinnat energiantuotannossa (ei sis. alv:a)". *In English: National fuel prices for energy generation purposes (VAT excluded).* Available only in Finnish. StatFin Database. Updated September 12th, 2018. Available at: https://www.stat.fi/tup/statfin/index.html

6. Suopajärvi, H. (2014) "Bioreducer use in blast furnace ironmaking in Finland: techno-economic assessment and CO_2 emission reduction potential". Dissertation. University of Oulu. Acta Universitatis Ouluensis, C Technica 513.

7. Suopajärvi, H., Umeki, K., Mousa, E., Hedayati, A., Romar, H., Kemppainen, A., Wang, C., Phounglamcheik, A., Tuomikoski, S., Norberg, N., Andefors, A., Öhman, M., Lassi, U. and Fabritius, T. (2018) "Use of biomass in integrated steelmaking – Status quo, future needs and comparison to other low-CO_2 steel production technologies" Applied Energy, 213, pp. 384–407. doi: 10.1016/j.apenergy.2018.01.060.

8. Toloue Farrokh, N., Sulasalmi, P. and Fabritius T. (2019) "Added Value for Forest Industry for Metals Producing and Processing Integrates (FOR&MET) – Project Report of University of Oulu". Available at: http://jultika.oulu.fi/Record/isbn978-952-62-2207-3

9. Valmet (2016) "Wingems". Available at: http://www.valmet.com/products/automation/solutions-for-pulp-and-paper/apc-advanced-process-controls/valmet-wingems/

10. VTT (2013) "Balas® Process Simulation Software". Available at: http://balas.vtt.fi/

11. Yamamoto, M. (2018) "St1 Cellunolix® process – Lignocellulosic bioethanol production and value chain upgrading", Bio4Fuels Days, October 12th 2018, Oslo.

575581 Conversion of Sugarcane Bagasse to Bioenergy over a Metal Oxide Catalyst

Valorization of sugarcane bagasse to high-quality biofuel precursors using metal oxide catalysts was studied in this work. Preliminary results showed that a hydrocarbon selectivity of 14.9 wt% can be achieved in the upgraded sugarcane bagasse derived pyrolysis oils. Non-condensed phenolic hydroxyl groups were almost eliminated in the upgraded pyrolysis oils. A detailed performance study of the metal oxide catalysts was carried out for the reaction condition optimization (temperature, time, catalyst loading). Different solvents (e.g. ethanol, methanol, isopropanol, acetone) were tested in this upgrading reaction. A series model compound study was conducted to study the possible repolymerization and condensation in this catalytic upgrading reaction. ^{1}H, ^{31}P, HSQC NMM, GPC, and GC-MS were used for pyrolysis oil structural analysis for understanding the reaction pathways. A stability test of the catalyst and a detailed characterization of the fresh/used catalysts were performed and was carried out.

577950 Fast Catalyst Evaluation Using Pyro-GC-MS for Lignin Conversion to Chemicals and Fuels

Lignin is a byproduct from wood pulping and cellulosic ethanol production. Currently most lignin is only burned for heat generation. Lignin has great potential to be one of the big drivers of switching from oil-based to bio-based fuels and chemicals. This paper presents a fast catalyst evaluation method for converting lignin to aromatics and liquid fuels using pyro-GC-MS. Six different catalysts, including ZSM-5, HY, FCC catalyst, Ru/alumina, Ni/beta, and Ru-Cu/HY catalyst which had been successfully applied in one-pot lignin conversion to jet fuel [1,2], were tested. For each catalyst, two temperatures, 600°C and 800°C, were compared. The lignin-derived intermediates, including toluene, xylenes, and phenols, were characterized by GC-MS. The detailed side-by-side comparison of the catalysts will be presented.

[1] H. Wang, H. Ruan, M. Feng, et al, One-Pot Process for Hydrodeoxygenation of Lignin to Alkanes Using Ru-Based Bimetallic and Bifunctional Catalysts Supported on Zeolite Y, ChemSusChem 2017, 10(8), 1846-1856.
[2] H. Wang, M. Feng, B. Yang, Catalytic hydrodeoxygenation of anisole: an insight into the role of metals in transalkylation reactions in bio-oil upgrading, Green Chemistry 2017, 19, 1668-1673.

557694 Molten Alkali Carbonates Pyrolysis of Lignin in Digestate for Phenolic Production

Molten alkali carbonates (MC) pyrolysis of digestate to produce phenolics was proposed in this work. The results showed that temperature strongly affected phenolic production by promoting the depolymerization of lignin, while particle size had less influence on phenolic productions owing to the catalysis and dynamic diffusion of alkali metal ions. Meanwhile, the effect of mass ratio between sample and molten carbonates was investigated by a series of proportions from 5 to 25 %. The optimum reaction conditions were 10 % of mass ratio in a particle size from 0.6 to 0.85 mm at 450 °C, with the maximum phenolics yield of 67.08 % in organic. In sum, 5.6 g of phenolics and 0.96 g of guaiacols were generated from 100 g of digestate. Compared with conventional pyrolysis method, the methoxyl group in liquid product could be efficiently removed at atmosphere in molten carbonates. Additionally, thermogravimetric analysis further demonstrated that alkali metal ions (Li^+, Na^+ and K^+) promoted the primary degradation of lignin and secondary monomolecular dissociation reactions of methoxyphenol.

588861 The Chemical Engineering Profession and Its Infinite Possibilities

Tayo Femi-Fowode is a passionate Chemical Engineer with over 10 years of corporate and leadership experience. He is currently a Business Integration Leader for GE Power's Monitoring and Diagnostic Center in Atlanta, GA. In his role, Tayo drives digital integration of 5000 monitored turbines, generators and other equipment to minimize downtime, optimize plant performance and save operating cost for more than 950 power plants in 75 countries and serving 350 million people worldwide. Before joining GE Power in 2011, he was Project Manager in the asset management division of Rolls-Royce Energy.

A second-generation Nigerian-American, Tayo holds an MBA from Columbia Business School with concentration in Strategy and Finance. He also holds a Master of Science and a Bachelor of Science *(magna cum laude)* degrees in Chemical Engineering from Columbia University and Prairie View A&M University respectively.

588862 Direct contact membrane distillation (DCMD) membrane distillation techniques for desalination

Direct contact membrane distillation (DCMD) is one of the simplest and most widely used membrane distillation techniques for desalination. It is a thermally driven separation process and is sought to achieve high water vapor flux while being thermally efficient. Typically, a hydrophobic porous membrane is used and water vapor from the salt solution is allowed to pass through the pores and condense on the other side of the membrane. Investigation on other structures of membranes is carried out. Membranes include those in the form of an integral composite, in which the structures of PVDF membranes are changed, and a stacked composite, in which different membranes are stacked on top of each other (e.g. a hydrophobic ePTFE membrane over a hydrophilic PVDF membrane, and a hydrophobic PVDF membrane over another hydrophobic PVDF membrane). These membranes were found to yield a higher water vapor flux than the conventional single hydrophobic membrane. Membranes are also characterized using a variety of characterization techniques: Contact angle, liquid entry pressure (LEP), bubble-point pressure, scanning electron microscopy (SEM), and atomic force microscopy (AFM).

588863 My Career Path

My career thus far has been shaped by many factors learned early in life. It started in my hometown of Yazoo City, Mississippi. It was there that I had the job that I consider to be my first engineering experience which was working with my dad at Gainwell Tire Services. While there, I discovered that I had a fondness for hands-on work and gravitated toward working closely with people as part of a team. This was instrumental in not only my early career development but shaped my life as a whole. From there, while in high school, I worked at our local construction company, Parker Builders. During this time, I developed a strong partiality for repairing process equipment. Also, a sense of work ethic was formed as there were several days where I worked on a hot tin roof for hours. After graduation, I headed to college at Louisiana Tech University. While there, I was a part of the football team while pursuing my Chemical Engineering degree. This conceivably impossible task was made possible largely due to the work ethic learned early in life. It also helped further developed my team-membership skills.

Later, I transferred to Southern Arkansas University. It was here that I not only compounded upon my team-membership skills but developed into a team leader as a captain of the football team. While there I took on several internships which included stints in the chemical, aerospace, energy, and mechanical fields. One of those internships was Lanxess Solutions in El Dorado, Arkansas. The skills learned early in life helped me very much in many of the projects given during this time and were good enough to be hired full time as a Shift Leader upon graduating college. In this role, I developed many technical skills and gained an understanding of the operations side of the plant. After some time there, I went to New Orleans, Louisiana to work at Galata Chemicals as a Process Engineer. While at Galata, I was able to further develop technical skills as well as learn the safety aspect that is imperative in a plant setting. After spending a year and a half there, I was offered a Process Engineer II position at Oxea Chemicals, my current role. Thus far I have learned even more about the chemical field and continue to grow professionally with the organizational training offered.

588932 Sustainable Energy Drivers: A Perspective from a Career in the Nuclear Sector

Dr. Harold T. Conner, Jr. has a 52-year career in the nuclear sector, having determined at an early age that he wanted a career in chemical engineering. Along the way, he blazed an impressive path by achieving a lot of firsts: the first African American student in the University of Tennessee-Martin Engineering School, the first cooperative engineering student at the University, and the first African American to receive bachelor's and master's degrees in chemical engineering from the University of Tennessee-Knoxville. He worked for more than 30 years in Oak Ridge, working in almost every facility at the Oak Ridge K-25 Site, a uranium enrichment facility that would be closed down in the mid-1980s and then undergo cleanup. Conner became site manager of the K-25 Site and was then named vice president of Environmental Management and Enrichment Facilities. He would leave Oak Ridge and worked at various DOE complexes across the nation until returning to Oak Ridge to help with the cleanup of the K-25 Site, which is now called the East Tennessee Technology Park (ETTP). Working for UCOR, an AECOM led partnership with Jacobs, his expertise has significantly helped the company work toward achieving Vision 2020, DOE's plan to complete ETTP cleanup by 2020. The site is being transformed into a private sector industrial park, national park, and conservation area. Conner is now semi-retired but serves as a consultant for UCOR.

588933 A Chemical Engineer in the Energy Journey

Dr. Quinta Nwanosike Warren, PE, PMP is a chemical engineer and an energy professional. She is an entrepreneur and has worked in industry, academia, and government. She holds a PhD in Chemical & BioMolecular Engineering from the Georgia Institute of Technology, and a Bachelors in Chemical Engineering from the Pennsylvania State University. She is a Project Management Professional, and a registered Professional Engineer.

Dr. Warren previously worked for ConocoPhillips for on CO2 capture research, and as a reservoir engineer for heavy oil and tight gas assets. She has worked with the US Department of Energy, learning about and contributing to domestic and international policy around carbon management and power generation. She also worked for the Millennium Challenge Corporation (MCC), a US government agency that works on international development. At MCC, she provided energy expertise and contributed to the development of energy projects.

Dr. Warren owns and runs a consulting company called Energy Research Consulting (www.EngrRC.com) which provides energy and water solutions and strategies. The company's projects include creating the STEM-focused Dakar American University of Science and Technology (www.DAUST.org) in Senegal, and designing a water treatment system for a mushroom farming initiative in Nigeria. Dr. Warren recently published a textbook titled "A Practical Guide to Oil & Gas Resource Characterization for Geologists and Reservoir Engineers."

578392 Scalable Production of Nanostructured Electrocatalysts for the e-Refinery

There is a broad consensus that the chemical industry will have to make a transition to renewable feedstocks. An important route will be to use electrochemical processes, driven by electricity from sustainable sources, to produce syngas, ammonia, and organic base chemicals from water, carbon dioxide, nitrogen, and/or biomass. In recent years, much attention has been given to developing and understanding novel electrocatalysts for this task, while much less efforts has yet been put in developing the corresponding reactor designs. That is why we started recently at Delft University of Technology the e-Refinery initiative: to make progress in the whole chain, from the molecular scale to large-scale system integration.

The materials used in electrochemical catalytic processes have yet to be designed and optimised. We aim at a rational design approach of the key component materials. Moreover, much attention will be given at the combination of high precision and scalability in the manufacturing of electrocatalysts and other materials. While currently most catalysts are made via liquid-phase processes with limited control, we use gas-phase production techniques such as atomic layer deposition to make our materials. This approach gives a high control over size distribution [1] and the composition [2] of the catalyst particles. Moreover, it has the potential to be applied in large-scale applications [3] making it an enabler for large-scale electrocatalytic processes.

In this presentation, I will show how our progress in the development of the e-Refinery concept in general, and the use in it of advanced nanoscale manufacturing techniques for the required materials in particular.

[1] Grillo, F., Van Bui, H., Moulijn, J. A., Kreutzer, M. T., & Van Ommen, J. R. (2017). Understanding and controlling the aggregative growth of platinum nanoparticles in atomic layer deposition: An avenue to size selection. The Journal of Physical Chemistry Letters, 8(5), 975-983.

[2] Cheng, N., Shao, Y., Liu, J., & Sun, X. (2016). Electrocatalysts by atomic layer deposition for fuel cell applications. Nano Energy, 29, 220-242.

[3] Van Bui, H., Grillo, F., & Van Ommen, J. R. (2017). Atomic and molecular layer deposition: off the beaten track. Chemical Communications, 53(1), 45-71.

575639 Machine Learning of Carbon Electrodes for Electric Double Layer Capacitors

Electric double layer capacitors (EDLCs) (a.k.a. supercapacitors or ultracapacitors) have attracted great research interest over the past few decades because of their potential applications for electrical energy storage with high power density, long-term cycle stability, and product safety. However, the energy density of EDLC is typically smaller than that of an alternative energy-storage device such as Li-ion battery, pressing for further improvements by using better electrode materials and charge carriers. With extremely high porosity and specific surface area, porous carbon shows great promise to lift the upper limit of the energy densities of EDLCs, thereby expanding their practical usage. Despite a wealth of literature on experimental explorations, it is still unclear how the structure and microscopic features of carbon electrodes affect the EDLC performance. In this work, we use machine learning methods to establish quantitative correlations between the structural features of porous carbons and their performance (i.e., energy and power density) based on extensive experimental data from the literature. We demonstrate that machine learning is able to not only reproduce the structure-properties relationships of EDLCs at static conditions but also forecast the device performance far away from equilibrium (e.g., charging/discharging behavior at high-scan rate). Moreover, we have identified structural features of carbon electrodes that can expand the region of supercapacitor performance in the Ragone plot.

577358 Applications of a-Nickel Hydroxide Catalyst for Electrochemical Water Splitting

Nickel hydroxide is a highly effective material for several electrochemical applications. Especially the disordered a-$Ni(OH)_2$ polymorph shows excellent performance for energy related applications. This work will cover few demonstrations of a-$Ni(OH)_2$ electrodeposits specifically geared towards catalysis of oxygen evolution reaction (OER) in alkaline electrochemical water splitting. Three different $Ni(OH)_2$-based nanostructured deposits fabricated using electric field directed deposition techniques will be explored. Standalone $Ni(OH)_2$ based deposits, as well as, stainless steel nanoparticle and carbon nanotube (CNT) supported $Ni(OH)_2$ deposits will be shown to be highly active OER catalysts. The standalone $Ni(OH)_2$ nanocarpets show remarkable bubble repellent behavior allowing durable and efficient catalysis. OER overpotentials as low as 450 mV to sustain 500 mA/cm^2 could be obtained with $Ni(OH)_2$ modified stainless steel deposits. The $Ni(OH)_2$ decorated CNT deposits can be further enhanced with platinum to produce excellent hydrogen evolution reaction catalysis. The OER/HER electrode combination shows stable water splitting at 10 mA/cm^2 for 24 hours. Effective strategies toward exploiting excellent $Ni(OH)_2$ activity via binder-free electrodeposits will be demonstrated.

Journal of Materials Chemistry A

ROYAL SOCIETY OF CHEMISTRY

COMMUNICATION

View Article Online
View Journal | View Issue

Check for updates

Cite this: J. Mater. Chem. A, 2019, 7, 965

Received 22nd November 2018
Accepted 7th December 2018

DOI: 10.1039/c8ta11223a

rsc.li/materials-a

A strongly coupled 3D ternary Fe_2O_3@Ni_2P/$Ni(PO_3)_2$ hybrid for enhanced electrocatalytic oxygen evolution at ultra-high current densities†

Xiaodi Cheng,[a] Zhiyan Pan, [b] Chaojun Lei,[a] Yangjun Jin,[b] Bin Yang,[a] Zhongjian Li, [a] Xingwang Zhang,[a] Lecheng Lei,[a] Chris Yuan [c] and Yang Hou [*a]

Developing highly efficient non-noble metal electrocatalysts for the oxygen evolution reaction (OER) at large current densities remains a challenge for water electrolysis. Herein, we developed a novel 3D ternary hybrid electrocatalyst consisting of Fe_2O_3 strongly coupled with Ni_2P/$Ni(PO_3)_2$ grown on commercial Ni foam (denoted as Fe_2O_3@Ni_2P/$Ni(PO_3)_2$/NF) through a thermal phosphorization reaction. In this 3D hybrid structure, the Fe_2O_3@Ni_2P/$Ni(PO_3)_2$ nanoparticles with a length of ~300 nm and width of ~200 nm are homogeneously grown on the surface of 3D Ni foam. Owing to the synergistic effect between Fe_2O_3@Ni_2P/$Ni(PO_3)_2$/NF and their strong coupling effect, the 3D Fe_2O_3@Ni_2P/$Ni(PO_3)_2$/NF hybrid achieves a superior electrocatalytic OER performance at ultra-high current densities in alkaline electrolyte. High catalytic current densities of 500 and 1000 mA cm^{-2} are reached by applying only potentials of 1.57 and 1.60 V, respectively, which are almost the lowest potentials among all previously reported transition metal (Ni and Fe) based phosphide and phosphate electrocatalysts, and even better than that of the benchmark Ir/C catalyst (1.64 and >1.8 V at 100 and 500 mA cm^{-2}). In situ Raman spectroscopy revealed that the NiOOH and FeOOH species were the real active phases in Fe_2O_3@Ni_2P/$Ni(PO_3)_2$/NF during the OER process. A two-electrode electrolyzer with Fe_2O_3@Ni_2P/$Ni(PO_3)_2$/NF delivers a cell voltage of 1.93, 2.48, and 3.02 V at 100, 500, and 1000 mA cm^{-2} for overall-water-splitting.

Electrochemical water splitting can transform renewable energies such as solar and wind into hydrogen fuels and hence is widely studied as a promising technology to address the possible energy crisis from increasing fossil fuel depletion in the future.[1-6] However, for the two half reactions in electrochemical water splitting, the four-step proton-coupled electron transfer in the oxygen evolution reaction (OER) process brings in a huge kinetic barrier which hampers the whole reaction rate.[7] Although it has been demonstrated that some noble metal-based materials (*e.g.*, IrO_2, RuO_2, and Ir/C) can work as highly efficient electrocatalysts to enhance the OER process, they still require a high potential of ~1.5 V to attain a tiny current density of 10 mA cm^{-2} in basic electrolyte. In addition, the high cost and scarcity of these noble metals also hinder their wide applications.[8-10] In recent years, research and development of low-cost, earth-abundant, and highly effective OER electrocatalysts such as transition metal sulfides,[11-16] nitrides,[17-19] borides,[20,21] carbides,[19,22,23] oxides,[24,25] hydroxides,[26-28] and oxyhydroxides[29,30] have attracted enormous attention and effort. Among transition metal based electrocatalysts, transition metal phosphides/phosphates have been found to be the most promising electrocatalysts with extreme activity and superior stability to catalyse the OER process in alkaline media.[31-34] Nonetheless, till now their catalytic performances are still far below the expected level for commercialization that needs a large current density >1000 mA cm^{-2} with high stability in alkaline solutions.[35,36]

Further research efforts are still needed to enhance the activity and stability of these transition-metal based electrocatalysts to facilitate their commercial application potential. It has been reported that construction of hybrid electrocatalysts based on transition metal phosphides/phosphates can synergize the advantages of the active components in the hybrid and can satisfy the prerequisites of high activity and stability for the OER.[37,38] In this regard, hybrids of transition metal oxides featuring iron element could be a promising OER electrocatalyst thanks to its inherent catalytic activity and natural abundance.

Herein, we develop a novel 3D strongly coupled ternary hybrid composed of Ni_2P/$Ni(PO_3)_2$ and Fe_2O_3 grown on 3D Ni foam through a disproportionation reaction. In such a hybrid system, the Fe_2O_3@Ni_2P/$Ni(PO_3)_2$ nanoparticles with an ~200 nm lateral size and ~300 nm transverse size were orderly

[a] Key Laboratory of Biomass Chemical Engineering of Ministry of Education, College of Chemical and Biological Engineering, Zhejiang University, Hangzhou, China. E-mail: yhou@zju.edu.cn

[b] Environmental Chemical and Resource Research Institute, College of Environment, Zhejiang University of Technology, Hangzhou, China

[c] Department of Mechanical and Aerospace Engineering, Case Western Reserve University, Cleveland, OH, 44106, USA

† Electronic supplementary information (ESI) available. See DOI: 10.1039/c8ta11223a

grown on the Ni foam. The 3D Fe_2O_3@$Ni_2P/Ni(PO_3)_2$/NF hybrid showed an excellent activity and robust stability for the OER in 1.0 M KOH, achieving extremely low potentials of 1.57 and 1.60 V at ultra-high current densities of 500 and 1000 mA cm^{-2}, respectively, which are practically the lowest potentials among all previously reported transition metal (Ni and Fe) based phosphide/phosphate OER electrocatalysts, and even exceed that of the state-of-the-art Ir/C catalyst. The outstanding OER activity can be attributed to the synergetic effect and the strong coupling effect between the Fe_2O_3 and $Ni_2P/Ni(PO_3)_2$. In situ Raman spectroscopy revealed that the NiOOH and FeOOH phases were the real active species in the Fe_2O_3@$Ni_2P/Ni(PO_3)_2$/NF for the OER. In addition to the admirable OER activity and stability, the bifunctional 3D Fe_2O_3@$Ni_2P/Ni(PO_3)_2$/NF delivered a cell voltage of 1.93, 2.48, and 3.02 V at 100, 500, and 1000 mA cm^{-2} toward overall-water-splitting, respectively.

The fabrication process of the 3D Fe_2O_3@$Ni_2P/Ni(PO_3)_2$/NF hybrid was based on two steps: firstly, a piece of ready-made Ni foam was dipped into $Fe(NO_3)_3$ solution and dried spontaneously in fume hood. Secondly, the resulting $Fe(NO_3)_3$/Ni foam was thermally phosphated at 300 °C for 1 h using red phosphorus sealed in a vacuum quartz tube to obtain the 3D Fe_2O_3@$Ni_2P/Ni(PO_3)_2$/NF hybrid. A possible mechanism towards the growth of the 3D Fe_2O_3@$Ni_2P/Ni(PO_3)_2$/NF hybrid was unveiled through two specific ingenious disproportionation reactions, with details provided in the ESI.† We methodically found out the effects of different phosphorus contents to acquire the highest OER capacity, and the red phosphorus amount of 100 mg was optimized (Fig. S1†).

The field-emission scanning electron microscopy (FESEM) images presented that the Fe_2O_3@$Ni_2P/Ni(PO_3)_2$ consists of particles with a width of ~200 nm and an average length of ~300 nm are orderly anchored on the surface of 3D Ni foam (Fig. 1a and b). The spatial skeleton morphology of the 3D Fe_2O_3@$Ni_2P/Ni(PO_3)_2$/NF hybrid remained undestroyed after the thermal phosphorization process (Fig. 1a and S2†). The elemental mapping images obtained *via* energy-dispersive X-ray spectroscopy (EDX) exhibited that the Ni, Fe, P and O elements were consistently distributed in the 3D Fe_2O_3@$Ni_2P/Ni(PO_3)_2$/NF hybrid (Fig. 1c). Transmission electron microscope (TEM) and high-resolution TEM (HRTEM) images of the 3D Fe_2O_3@$Ni_2P/Ni(PO_3)_2$ hybrid showed that the Fe_2O_3 and $Ni_2P/Ni(PO_3)_2$

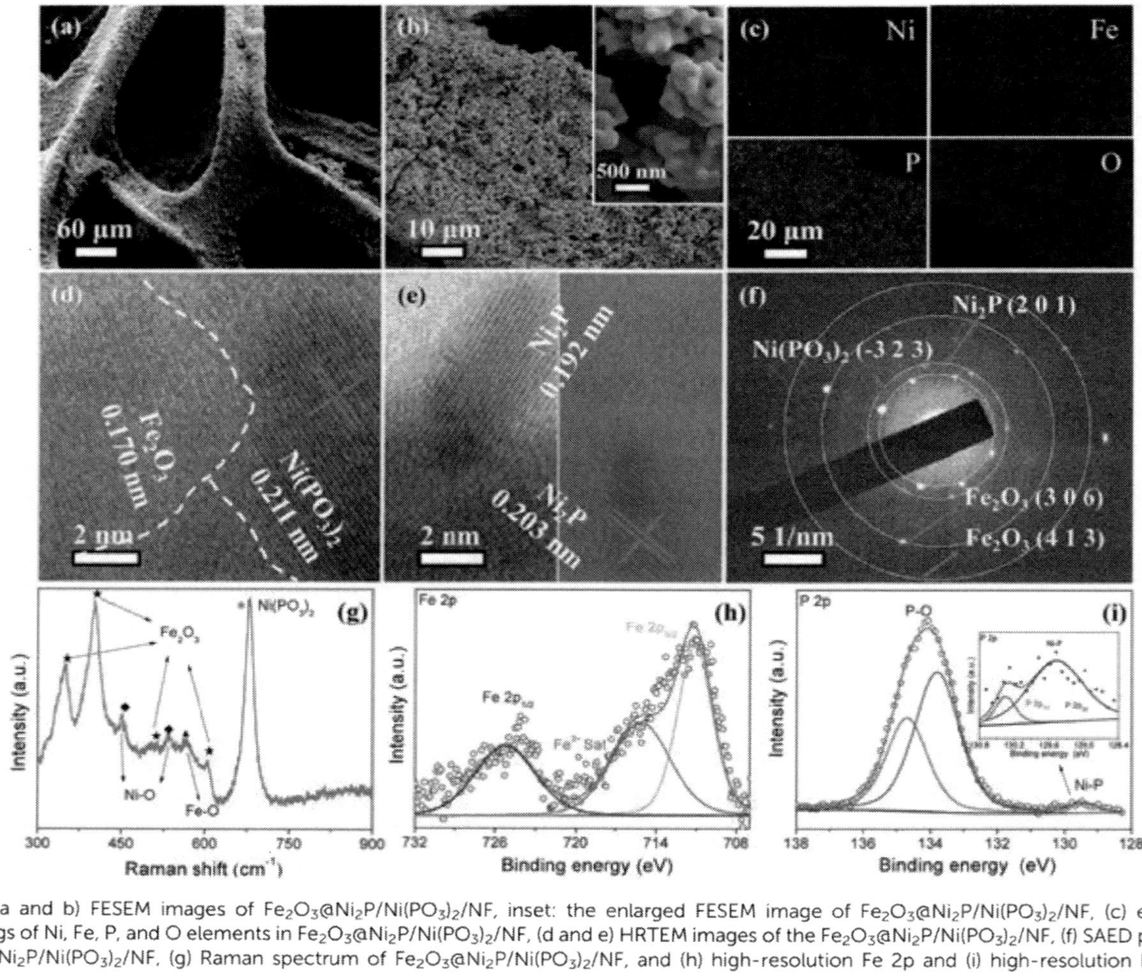

Fig. 1 (a and b) FESEM images of Fe_2O_3@$Ni_2P/Ni(PO_3)_2$/NF, inset: the enlarged FESEM image of Fe_2O_3@$Ni_2P/Ni(PO_3)_2$/NF, (c) elemental mappings of Ni, Fe, P, and O elements in Fe_2O_3@$Ni_2P/Ni(PO_3)_2$/NF, (d and e) HRTEM images of the Fe_2O_3@$Ni_2P/Ni(PO_3)_2$/NF, (f) SAED pattern of Fe_2O_3@$Ni_2P/Ni(PO_3)_2$/NF, (g) Raman spectrum of Fe_2O_3@$Ni_2P/Ni(PO_3)_2$/NF, and (h) high-resolution Fe 2p and (i) high-resolution P 2p XPS were successfully formed on the Ni foam (Fig. 1d, e and S3†). An obvious boundary of the crystal surface of the Fe_2O_3 and $Ni_2P/Ni(PO_3)_2$ was observed. The $Fe_2O_3@Ni_2P/Ni(PO_3)_2$/NF hybrid has interplanar spacings of 0.170 and 0.211 nm, attributed to the (116) plane of Fe_2O_3 and (−224) plane of $Ni(PO_3)_2$, respectively. The other two lattice spacings of 0.192 and 0.203 nm correspond to the lattice planes (210) and (201) of Ni_2P, respectively. The selected-area electron diffraction (SAED) pattern showed the highly crystalline nature of Fe_2O_3, $Ni(PO_3)_2$, and Ni_2P in the 3D $Fe_2O_3@Ni_2P/Ni(PO_3)_2$/NF hybrid (Fig. 1f).

X-ray diffraction (XRD) pattern of the 3D $Fe_2O_3@Ni_2P/Ni(PO_3)_2$/NF hybrid confirmed the coexistence of Fe_2O_3, Ni_2P, and $Ni(PO_3)_2$ (Fig. S4†). Furthermore, the Raman spectrum of the 3D $Fe_2O_3@Ni_2P/Ni(PO_3)_2$/NF hybrid exhibited four prominent peaks located at 349, 403, 512, and 605 cm^{-1}, corresponding to the vibrations of Fe_2O_3 (Fig. 1g).[39,40] The peak located at 678 cm^{-1} was detected in the stretching vibration modes of symmetric PO_2^-, indicating the presence of $Ni(PO_3)_2$. Although no Raman peaks of Ni_2P are detected under the condition of a specific excitation laser,[1] two observed peaks located at 453 and 534 cm^{-1} for the 3D $Fe_2O_3@Ni_2P/Ni(PO_3)_2$/NF hybrid could be attributed to the Ni–O bonds because of the oxidation species of Ni_2P. The X-ray photoelectron spectroscopy (XPS) survey spectrum of the 3D $Fe_2O_3@Ni_2P/Ni(PO_3)_2$/NF hybrid revealed the existence of Fe, Ni, P, and O elements (Fig. S5†). The high resolution Fe 2p XPS spectra of the $Fe_2O_3@Ni_2P/Ni(PO_3)_2$/NF hybrid showed three principal peaks at 725.1, 715.0, and 711.0 eV in Fe 2p (Fig. 1h), which can be attributed to the Fe $2p_{1/2}$, Fe^{3+} satellite, and Fe $2p_{3/2}$, respectively, confirming the formation of Fe_2O_3.[41] In the high-resolution P 2p XPS spectra of 3D $Fe_2O_3@Ni_2P/Ni(PO_3)_2$/NF (Fig. 1i), the peaks centered at 134.7 eV and 138.8 eV are assigned to the P–O species, revealing the existence of the $Ni(PO_3)_2$ component.[1] The peaks at 130.3 and 129.4 eV were attributed to the Ni–P species present, confirming the presence of Ni_2P. The lower peak shifted negatively which could be attributed to the partial electrons from surrounding metallic Ni accepted by P species, which validates the strongly coupled effects in the $Fe_2O_3@Ni_2P/Ni(PO_3)_2$/NF hybrid.[42] The high-resolution O 1s XPS spectra of the 3D $Fe_2O_3@Ni_2P/Ni(PO_3)_2$/NF hybrid showed two peaks located at 533.3 and 531.7 eV corresponding to H–O–H and Fe(Ni)–O bonds, due to the adsorbed H_2O, Fe_2O_3, and $Ni(PO_3)_2$, respectively (Fig. S6†).[1] All the above results proved the successful formation of $Ni_2P/Ni(PO_3)_2$ with Fe_2O_3 incorporated into the 3D $Fe_2O_3@Ni_2P/Ni(PO_3)_2$/NF hybrid.

The OER electrocatalytic activity of the 3D $Fe_2O_3@Ni_2P/Ni(PO_3)_2$/NF hybrid was first investigated in 1.0 M KOH electrolyte in a three-electrode cell. As a contrast, Ni_2P/NF, Fe_2O_3/NF, and commercial Ir/C/NF were also tested (Fig. 2a). The potential reported here was switched to the value *versus* the reversible hydrogen electrode (RHE). The 3D $Fe_2O_3@Ni_2P/Ni(PO_3)_2$/NF hybrid only required low potentials of 1.53, 1.57, and 1.60 V to realize the ultra-high current densities of 100, 500, and 1000 mA cm^{-2}, respectively. Notably, a high current density of 1200 mA cm^{-2} (>1000 mA cm^{-2}) can be achieved at a relative low potential of 1.61 V, which makes the 3D $Fe_2O_3@Ni_2P/Ni(PO_3)_2$/NF hybrid a promising OER electrocatalyst towards practical application in industrial water splitting in basic electrolyte.[35,36] The 3D $Fe_2O_3@Ni_2P/Ni(PO_3)_2$/NF hybrid showed a much higher OER catalytic activity than the Ni_2P/NF with potentials of 1.55, 1.63, and 1.72 V at 100, 500, and 1000 mA cm^{-2} and Fe_2O_3/NF with potentials of 1.54, 1.60, and 1.66 V at 100, 500, and 1000 mA cm^{-2}, respectively. Comparison of the

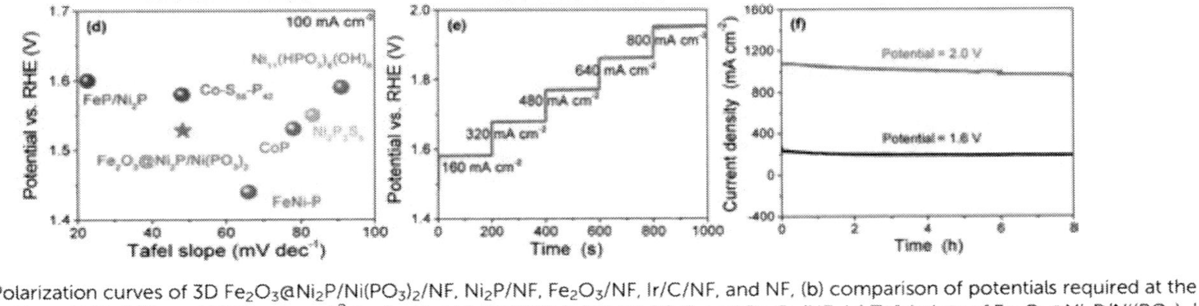

Fig. 2 (a) Polarization curves of 3D $Fe_2O_3@Ni_2P/Ni(PO_3)_2$/NF, Ni_2P/NF, Fe_2O_3/NF, Ir/C/NF, and NF, (b) comparison of potentials required at the current densities of 100, 500, and 1000 mA cm^{-2} for $Fe_2O_3@Ni_2P/Ni(PO_3)_2$/NF, Ni_2P/NF, and Fe_2O_3/NF, (c) Tafel plots of $Fe_2O_3@Ni_2P/Ni(PO_3)_2$/NF, Ni_2P/NF, Fe_2O_3/NF, and Ir/C/NF, (d) comparison of Tafel slopes and potentials required for a current density of 100 mA cm^{-2}, (e) multiple-current step curve of $Fe_2O_3@Ni_2P/Ni(PO_3)_2$/NF without *iR*-correction, (f) chronopotentiometric measurements of $Fe_2O_3@Ni_2P/Ni(PO_3)_2$/NF at potentials of 1.6 and 2.0 V without *iR*-correction. All the above results were obtained in 1.0 M KOH.

OER performance among the 3D Fe_2O_3@$Ni_2P/Ni(PO_3)_2$/NF hybrid, Ni_2P/NF, and Fe_2O_3/NF confirms the existence of a synergistic effect among Fe_2O_3, Ni_2P, and $Ni(PO_3)_2$ components (Fig. 2b), which contributes to the high OER activity of 3D Fe_2O_3@$Ni_2P/Ni(PO_3)_2$/NF. Such low potentials of the 3D Fe_2O_3@$Ni_2P/Ni(PO_3)_2$/NF hybrid was even smaller than that of the commercial Ir/C/NF electrocatalyst (potential of 1.64 and >1.8 V at 100 and 500 mA cm^{-2}).

The Tafel analysis was used to investigate the OER kinetics (Fig. 2c). A lower Tafel slope of 48.2 mV dec^{-1} was observed for the 3D Fe_2O_3@$Ni_2P/Ni(PO_3)_2$/NF hybrid in comparison to 69.8 mV dec^{-1} for Ni_2P/NF, 73.8 mV dec^{-1} for Fe_2O_3/NF, and 94.0 mV dec^{-1} for Ir/C/NF, indicating that the 3D Fe_2O_3@$Ni_2P/Ni(PO_3)_2$/NF hybrid possesses a favorable reaction kinetics toward the OER. The potential and corresponding Tafel slope of the 3D Fe_2O_3@$Ni_2P/Ni(PO_3)_2$/NF hybrid are almost the lowest potential at a current density of 100 mA cm^{-2} and nearly the lowest Tafel slope among all previously reported transition metal (Ni and Fe) based phosphide and phosphate electrocatalysts (Fig. 2d and Table S1†),[35,43–47] which further demonstrates the excellent OER performance of the 3D Fe_2O_3@$Ni_2P/Ni(PO_3)_2$/NF hybrid. The multiple-current step curve of the 3D Fe_2O_3@$Ni_2P/Ni(PO_3)_2$/NF hybrid at increasing current densities from 160 to 800 mA cm^{-2} is shown in Fig. 2e. With the start current density of 160 mA cm^{-2}, the potential step located at 1.58 V remained unaffected for the rest 200 s. The other four steps also showed similar trends, denoting the exceptional mechanical robustness and excellent mass transportation by changing the cation species (Ni, Co, and Cu) of precursors of the 3D Fe_2O_3@$Ni_2P/Ni(PO_3)_2$/NF hybrid during the OER process. It is worth mentioning that the 3D Fe_2O_3@$Ni_2P/Ni(PO_3)_2$/NF hybrid exhibited a superior stability with an insignificant fading of current density of 200 and 1000 mA cm^{-2} at 1.6 and 2.0 V over 10 h of continuous OER (Fig. 2f). The FESEM image of the Fe_2O_3@$Ni_2P/Ni(PO_3)_2$/NF hybrid after the OER test showed slight changes in morphology (Fig. S7†). The corresponding XRD and XPS results are shown in Fig. S8–S10,† in which the Fe_2O_3 and $Ni_2P/Ni(PO_3)_2$ species in the 3D Fe_2O_3@$Ni_2P/Ni(PO_3)_2$ hybrid were fully converted into FeOOH and NiOOH species after a 10 h OER test, which is consistent with the operando Raman results in the following section and previously reported results.[35,36]

To identify the synergistic effect among Fe_2O_3, Ni_2P, and $Ni(PO_3)_2$ towards the extrusive OER performance, a series of electrochemical measurements were conducted. The electrochemical double layer capacitances (C_{dl}) of the 3D Fe_2O_3@$Ni_2P/Ni(PO_3)_2$/NF and Ni_2P/NF showed that the C_{dl} of 21.1.0 mF cm^{-2} for the 3D Fe_2O_3@$Ni_2P/Ni(PO_3)_2$/NF was much higher than 15.2 mF cm^{-2} for Ni_2P/NF (Fig. 3a and S11†), elucidating that the 3D Fe_2O_3@$Ni_2P/Ni(PO_3)_2$/NF hybrid possessed more active surface area compared with the Ni_2P/NF. The C_{dl} (21.1.0 mF cm^{-2}) of the 3D Fe_2O_3@$Ni_2P/Ni(PO_3)_2$/NF was slightly lower than that of Fe_2O_3/NF (22.2 mF cm^{-2}), which can be explained by the fact that the actual activity was determined by the number of real activity phases produced during the OER process.[48] The electrochemical impedance spectroscopy (EIS) results are shown in Fig. 3b, and the 3D Fe_2O_3@$Ni_2P/Ni(PO_3)_2$/NF hybrid exhibited a much smaller charge-transfer resistance compared with the Ni_2P/NF and Fe_2O_3, suggesting a fast electron transfer in the 3D

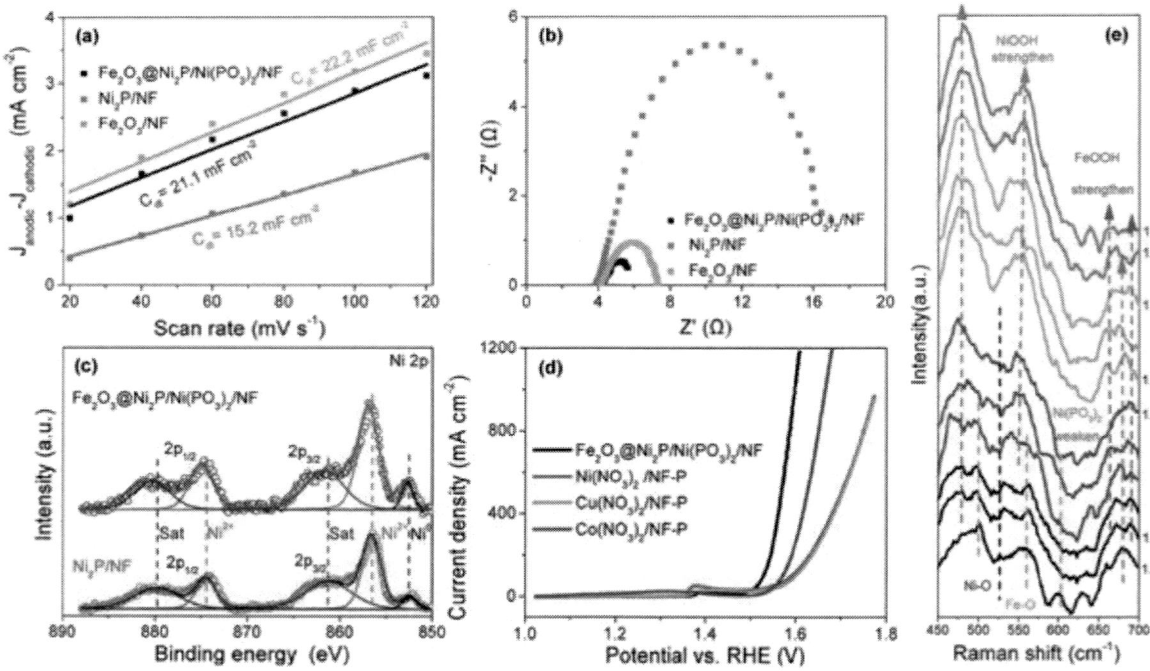

Fig. 3 (a) ECSAs of Fe_2O_3@$Ni_2P/Ni(PO_3)_2$/NF, Ni_2P/NF, and Fe_2O_3/NF, (b) Nyquist plots of Fe_2O_3@$Ni_2P/Ni(PO_3)_2$/NF, Ni_2P/NF, and Fe_2O_3/NF, (c) high-resolution Ni 2p XPS spectra of Fe_2O_3@$Ni_2P/Ni(PO_3)_2$/NF and Ni_2P/NF, (d) polarization curves of $M(NO_3)_x$/NF–P (M = Fe, Ni, Co, and Cu), and (e) operando Raman spectra of Fe_2O_3@$Ni_2P/Ni(PO_3)_2$/NF with different applied potentials.

Fe_2O_3@$Ni_2P/Ni(PO_3)_2$/NF hybrid. In addition, in order to investigate the strong coupling effect among Fe_2O_3, Ni_2P, and $Ni(PO_3)_2$ for the superior OER performance, additional XPS measurements were conducted. In the high-resolution Ni 2p XPS spectra (Fig. 3c), three peaks located at 875.0, 857.0, and 852.6 eV can be ascribed to the Ni^{2+} $2p_{1/2}$, Ni^{2+} $2p_{3/2}$, and Ni^0, respectively. The other peaks centered at 880.4 and 861.7 eV in the Ni 2p XPS spectra can be indexed to the two satellite peaks of Ni 2p. Compared with the Ni_2P/NF (874.4 eV for Ni^{2+} $2p_{1/2}$ and 856.5 eV for Ni^{2+} $2p_{3/2}$), the binding energy of 3D Fe_2O_3@$Ni_2P/Ni(PO_3)_2$/NF was positively shifted by about 0.6 eV, demonstrating the existence of a strong coupling effect.[49] The positive shifts of binding energy from Ni_2P/NF to Fe_2O_3@$Ni_2P/Ni(PO_3)_2$/NF could be attributed to the raised valence of Ni from Ni_2P to $Ni(PO_3)_2$.[37] These results demonstrated that the synergistic effect and strong coupling effect between Fe_2O_3 and $Ni_2P/Ni(PO_3)_2$ accounted for the exceptional OER performance of the 3D Fe_2O_3@$Ni_2P/Ni(PO_3)_2$/NF hybrid. Importantly, the thermal phosphorization strategy can be also extended to synthesize a series of transition metal oxide coupled nickel phosphide/phosphate hybrids (Fig. 3d), and the results showed that the 3D Fe_2O_3@$Ni_2P/Ni(PO_3)_2$/NF hybrid has the highest OER catalytic performance.

In order to further understand the transformation of the composition and local structure of the 3D Fe_2O_3@$Ni_2P/Ni(PO_3)_2$/NF hybrid during the OER process, *in situ* Raman measurements were conducted (Fig. 3e). The 3D Fe_2O_3@$Ni_2P/Ni(PO_3)_2$/NF hybrid exhibited a crystal structure with potentials of below 1.2 V for the OER. The $Ni(PO_3)_2$ peak located at 678 cm^{-1}, and the Fe_2O_3 peak located at 508 and 606 cm^{-1} did not change at low potentials below 1.30 V. The peak intensity of $Ni(PO_3)_2$ becomes weak with an irreversible transformation with the potentials >1.30 V. Meanwhile, this process facilitates structural conversion from the $Ni(PO_3)_2$ and Fe_2O_3 structures to complex amorphous components with peaks located at 479 and 555 cm^{-1} for NiOOH species, and 661 and 690 cm^{-1} for FeOOH species, respectively.[13,50] The Fe_2O_3 structure was practically converted into a complex FeOOH phase with the potentials >1.5 V, while the $Ni(PO_3)_2$ structure was completely converted to a complex NiOOH phase with the potentials >1.7 V. Based on the above results and previous reports,[27,51–53] one can conclude that the amorphous NiOOH and FeOOH structures as real active phases contributed to the high performance during the OER process.

Encouraged by the superior OER performance, the 3D Fe_2O_3@$Ni_2P/Ni(PO_3)_2$/NF hybrid as a bi-functional electrocatalyst for overall-water-splitting was examined in a two-electrode cell in 1.0 M KOH electrolyte. The 3D Fe_2O_3@$Ni_2P/Ni(PO_3)_2$/NF exhibited a bifunctional catalytic performance with the potentials of 1.93, 2.48, and 3.02 V to reach the ultra-high current

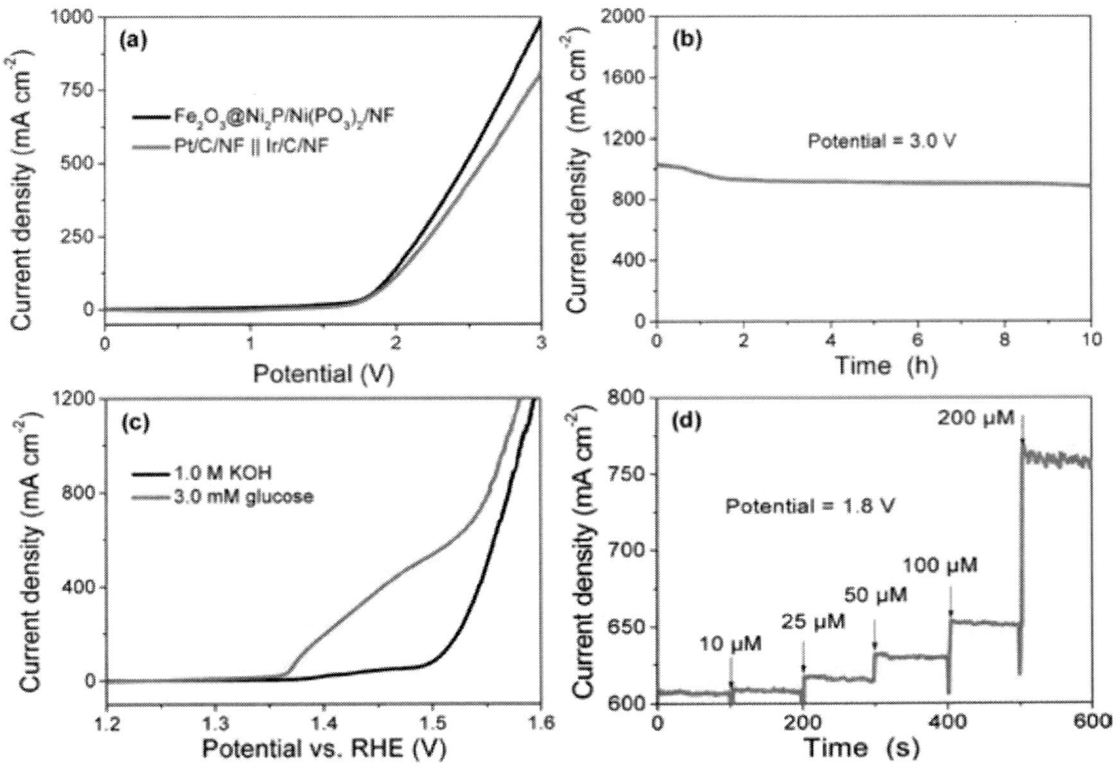

Fig. 4 (a) Polarization curves of bifunctional Fe_2O_3@$Ni_2P/Ni(PO_3)_2$/NF and Pt/C/NF–Ir/C/NF for overall-water-splitting, (b) chronoamperometric measurement at 3.0 V across Fe_2O_3@$Ni_2P/Ni(PO_3)_2$/NF without *iR*-correction, (c) polarization curves of Fe_2O_3@$Ni_2P/Ni(PO_3)_2$/NF in 1.0 M KOH and 3.0 mM glucose solution, (d) amperometric response of Fe_2O_3@$Ni_2P/Ni(PO_3)_2$ at 1.8 V to successive addition of glucose to 1.0 M KOH without *iR*-correction.

densities of 100, 500, and 1000 mA cm^{-2}, respectively. Notably, such a high OER performance even surpassed the benchmarking catalytic combination using commercial Pt/C as the cathode and commercial Ir/C as the anode with the current densities of 100 and 500 mA cm^{-2} attained at potentials of 1.97 and 2.59 V, respectively (Fig. 4a). The long-term robust stability of the 3D Fe_2O_3@Ni_2P/$Ni(PO_3)_2$/NF hybrid was confirmed with the applied current density remaining steady at 1000 mA cm^{-2} at a potential of 3.0 V for >10 h overall-water-splitting (Fig. 4b). Importantly, the 3D Fe_2O_3@Ni_2P/$Ni(PO_3)_2$/NF can be further extended to catalyze the glucose oxidation reaction. To investigate the capability of the 3D Fe_2O_3@Ni_2P/$Ni(PO_3)_2$/NF hybrid for glucose oxidation, we subsequently measured the polarization curves of 3D Fe_2O_3@Ni_2P/$Ni(PO_3)_2$/NF in 1.0 M KOH and 3.0 mM glucose mixture solution (Fig. 4c). After injection of 3.0 mM glucose, the Fe_2O_3@Ni_2P/$Ni(PO_3)_2$/NF showed a drastically higher anodic current density than that in 1.0 M KOH, suggesting the better electrocatalytic activity of the Fe_2O_3@Ni_2P/$Ni(PO_3)_2$/NF towards glucose oxidation. The conspicuous current densities increase were detected with per addition of glucose at applied potentials of 1.8 V (Fig. 4d), indicating that the Fe_2O_3@Ni_2P/$Ni(PO_3)_2$/NF possessed a fast and sensitive response to glucose oxidation.

Conclusion

In conclusion, a 3D strongly coupled ternary Fe_2O_3@Ni_2P/$Ni(PO_3)_2$/NF hybrid electrocatalyst was developed through an ingenious disproportionation reaction. The 3D Fe_2O_3@Ni_2P/$Ni(PO_3)_2$ hybrid delivered low potentials of 1.57 and 1.60 V at ultra-high current densities of 500 and 1000 mA cm^{-2}, respectively, towards the OER in basic media. Such potentials of the Fe_2O_3@Ni_2P/$Ni(PO_3)_2$ hybrid are almost the lowest among all previously reported transition metal (Ni and Fe) based nickel phosphide/nickel phosphate electrocatalysts, and even outperform the commercial Ir/C catalyst. The excellent OER catalytic activity of the 3D Fe_2O_3@Ni_2P/$Ni(PO_3)_2$/NF could be attributed to the synergic effect among the Fe_2O_3, Ni_2P, and $Ni(PO_3)_2$ in addition to their strong coupling effect. It was proved that the NiOOH and FeOOH species acted as the active phases in Fe_2O_3@Ni_2P/$Ni(PO_3)_2$/NF during the OER process. The developed Fe_2O_3@Ni_2P/$Ni(PO_3)_2$/NF presented in this work may inspire a way to synthesize transition metal oxides coupled with transition metal phosphide and phosphate hybrids toward energy conversion applications, such as hydrogen evolution reactions, nitrogen reduction reactions, and CO_2 reduction reactions.

Conflicts of interest

There are no conflicts to declare.

Acknowledgements

Y. Hou acknowledges the support of the NSFC 21878270 and 51702284 and the Startup Foundation for Hundred-Talent Program of Zhejiang University (112100-193820101/001/022).

Notes and references

1 H. Zhou, F. Yu, J. Sun, R. He, S. Chen, C. W. Chu and Z. Ren, *Proc. Natl. Acad. Sci. U. S. A.*, 2017, **114**, 5607–5611.
2 X. Zou and Y. Zhang, *Chem. Soc. Rev.*, 2015, **44**, 5148–5180.
3 C. Lei, H. Chen, J. Cao, J. Yang, M. Qiu, Y. Xia, C. Yuan, B. Yang, Z. Li, X. Zhang, L. Lei, J. Abbott, Y. Zhong, X. Xia, G. Wu, Q. He and Y. Hou, *Adv. Energy Mater.*, 2018, **8**, 1801912.
4 J. H. Cao, C. J. Lei, B. Yang, Z. J. Li, L. C. Lei, Y. Hou and X. L. Feng, *Batteries & Supercaps*, 2018, DOI: 10.1002/batt.201800098.
5 R. Gao, G. Yu, W. Chen, G. D. Li, S. Gao, Z. Zhang, X. Shen, X. Huang and X. Zou, *ACS Appl. Mater. Interfaces*, 2018, **10**, 696–703.
6 Y. Chen, G. Yu, W. Chen, Y. Liu, G. D. Li, P. Zhu, Q. Tao, Q. Li, J. Liu, X. Shen, H. Li, X. Huang, D. Wang, T. Asefa and X. Zou, *J. Am. Chem. Soc.*, 2017, **139**, 12370–12373.
7 H. Schäfer and M. Chatenet, *ACS Energy Lett.*, 2018, **3**, 574–591.
8 O. Kasian, J. P. Grote, S. Geiger, S. Cherevko and K. J. J. Mayrhofer, *Angew. Chem., Int. Ed.*, 2017, **57**, 2488–2491.
9 J. Luo, J. H. Im, M. T. Mayer, M. Schreier, M. K. Nazeeruddin, N. G. Park, S. D. Tilley, H. J. Fan and M. Grätzel, *Science*, 2014, **345**, 1593.
10 B. You and Y. Sun, *Acc. Chem. Res.*, 2018, **51**, 1571–1580.
11 Y. Liu, Q. Li, R. Si, G. D. Li, W. Li, D. P. Liu, D. Wang, L. Sun, Y. Zhang and X. Zou, *Adv. Mater.*, 2017, **29**, 1606200.
12 J. Hu, A. Liu, H. Jin, D. Ma, D. Yin, P. Ling, S. Wang, Z. Lin and J. Wang, *J. Am. Chem. Soc.*, 2015, **137**, 11004–11010.
13 X. Zou, Y. Liu, G. D. Li, Y. Wu, D. P. Liu, W. Li, H. W. Li, D. Wang, Y. Zhang and X. Zou, *Adv. Mater.*, 2017, **29**, 1700404.
14 J. Cao, C. Lei, J. Yang, X. Cheng, Z. Li, B. Yang, X. Zhang, L. Lei, Y. Hou and K. Ostrikov, *J. Am. Chem. Soc.*, 2018, **6**, 18877–18883.
15 H. Yang, Q. Ming, N. Gyutae, K. Min Gyu, Z. Tao, l. kejun, z. xiaodong, C. Jaephil, Y. Chris and F. Xinliang, *Nano Lett.*, 2017, **17**, 4202–4209.
16 Z. Xiaoxin, W. Yuanyuan, L. Yipu, L. Dapeng, L. Wang, G. Lin, L. Huan, W. Pengwei, S. Lei and Z. Yu, *Chem*, 2018, **4**, 1139–1152.
17 X. Li, J. Wei, Q. Li, S. Zheng, Y. Xu, P. Du, C. Chen, J. Zhao, H. Xue, Q. Xu and H. Pang, *Adv. Funct. Mater.*, 2018, **28**, 1800886.
18 S. Zhao, M. Li, M. Han, D. Xu, J. Yang, Y. Lin, N. E. Shi, Y. Lu, R. Yang, B. Liu, Z. Dai and J. Bao, *Adv. Funct. Mater.*, 2018, **28**, 1706018.
19 C. Lei, Y. Wang, Y. Hou, P. Liu, J. Yang, T. Zhang, X. Zhuang, M. Chen, B. Yang, L. Lei, C. Yuan, M. Qiu and X. Feng, *Energy Environ. Sci.*, 2018, DOI: 10.1039/c8ee01841c.
20 J. M. V. Nsanzimana, Y. Peng, Y. Y. Xu, L. Thia, C. Wang, B. Y. Xia and X. Wang, *Adv. Energy Mater.*, 2017, **8**, 1701475.
21 J. Masa, I. Sinev, H. Mistry, E. Ventosa, M. de la Mata, J. Arbiol, M. Muhler, B. Roldan Cuenya and W. Schuhmann, *Adv. Energy Mater.*, 2017, **7**, 1700381.

22 C. Zhihua, F. Qiang, H. Qing, X. Yukun, L. Yuan, Z. Yang and Q. Liangti, *Adv. Funct. Mater.*, 2018, **28**, 1705967.
23 S. Hao, L. Yuebin, Y. Cheng, X. Likun, Q. Pengwei, M. Qiaoqiao, Z. Xiaohui, G. Jun, D. Zhao and P. Yang, *Energy Environ. Sci.*, 2018, **11**, 2363–2371.
24 P. Cai, J. Huang, J. Chen and Z. Wen, *Angew. Chem., Int. Ed.*, 2017, **56**, 4858–4861.
25 Y. Yang, M. Luo, Y. Xing, S. Wang, W. Zhang, F. Lv, Y. Li, Y. Zhang, W. Wang and S. Guo, *Adv. Mater.*, 2018, **30**, 1706085.
26 A. Indra, U. Paik and T. Song, *Angew. Chem., Int. Ed.*, 2018, **57**, 1241–1245.
27 Z. Chen, L. Cai, X. Yang, C. Kronawitter, L. Guo, S. Shen and B. E. Koel, *ACS Catal.*, 2018, **8**, 1238–1247.
28 H. Yang, R. L. Martin, Z. Jian, L. Shaohua, Z. Xiaodong and F. Xinliang, *Energy Environ. Sci.*, 2016, **9**, 478–483.
29 H. J. Zhang, X. P. Li, A. Hahnel, V. Naumann, C. Lin, S. Azimi, S. L. Schweizer, A. W. Maijenburg and R. B. Wehrspohn, *Adv. Funct. Mater.*, 2018, **28**, 1706847.
30 G. Anandhababu, Y. Y. Huang, D. D. Babu, M. X. Wu and Y. B. Wang, *Adv. Funct. Mater.*, 2018, **28**, 1706120.
31 J. Xu, J. Li, D. Xiong, B. Zhang, Y. Liu, K. H. Wu, I. Amorim, W. Li and L. Liu, *Chem. Sci.*, 2018, **9**, 3470–3476.
32 P. F. Liu, X. Li, S. Yang, M. Y. Zu, P. Liu, B. Zhang, L. R. Zheng, H. Zhao and H. G. Yang, *ACS Energy Lett.*, 2017, **2**, 2257–2263.
33 J. Huang, Y. Sun, Y. Zhang, G. Zou, C. Yan, S. Cong, T. Lei, X. Dai, J. Guo, R. Lu, Y. Li and J. Xiong, *Adv. Mater.*, 2017, **30**, 1705045.
34 H. Yang, Q. Ming, Z. Tao, Z. Xiaodong, K. Chang Soo, Y. Chris and F. Xinliang, *Adv. Mater.*, 2017, **29**, 1604480.
35 F. Yu, H. Zhou, Y. Huang, J. Sun, F. Qin, J. Bao, W. A. Goddard 3rd, S. Chen and Z. Ren, *Nat. Commun.*, 2018, **9**, 2551.
36 Y. Liu, X. Liang, L. Gu, Y. Zhang, G. D. Li, X. Zou and J. S. Chen, *Nat. Commun.*, 2018, **9**, 2609.
37 Q. Liu, C. Chen, J. Zheng, L. Wang, Z. Yang and W. Yang, *J. Am. Chem. Soc.*, 2017, **5**, 1421–1427.
38 X. Junyuan, P. S. S. Juliana, E. M. Natalia, C. José Diogo, Y. P. Dmitri, K. Kirill, I. L. Oleg and V. K. k. Yury, *ACS Catal.*, 2018, **8**, 2595–2600.
39 I. Chourpa, L. Douziech Eyrolles, L. Ngaboni Okassa, J. F. Fouquenet, S. Cohen Jonathan, M. Souce, H. Marchais and P. Dubois, *Analyst*, 2005, **130**, 1395–1403.
40 H. Sandra, P. Dmitrii, P. Yuri, B. Sebastian, M. Alina and B. Julien, *ChemSusChem*, 2017, **10**, 3644–3651.
41 F. Zhang, Y. M. Shi, T. Xue, J. F. Zhang, Y. Liang and B. Zhang, *Sci. China Mater.*, 2017, **60**, 324–334.
42 L. Chen, L. Lu, H. Zhu, Y. Chen, Y. Huang, Y. Li and L. Wang, *Nat. Commun.*, 2017, **8**, 14136.
43 W. Li, X. Gao, D. Xiong, F. Xia, J. Liu, W. G. Song, J. Xu, S. M. Thalluri, M. F. Cerqueira, X. Fu and L. Liu, *Chem. Sci.*, 2017, **8**, 2952–2958.
44 Z. Dai, H. Geng, J. Wang, Y. Luo, B. Li, Y. Zong, J. Yang, Y. Guo, Y. Zheng, X. Wang and Q. Yan, *ACS Nano*, 2017, **11**, 11031–11040.
45 X. Zhang, S. Zhang, J. Li and E. Wang, *J. Mater. Chem. A*, 2017, **5**, 22131–22136.
46 B. Zhang, Y. H. Lui, H. Ni and S. Hu, *Nano Energy*, 2017, **38**, 553–560.
47 W. M. Prashanth, P. Chakadola, L. Stefan, B. B. Florian, W. Carsten, S. Michael, D. Xiaohui, D. Holger and D. Matthias, *Energy Environ. Sci.*, 2018, **11**, 1287–1298.
48 Y. Liu, Y. Su, X. Quan, X. Fan, S. Chen, H. Yu, H. Zhao, Y. Zhang and J. Zhao, *ACS Catal.*, 2018, **8**, 1186–1191.
49 H. Yang, W. Zhenhai, C. Shumao, F. Xinliang and C. Junhong, *Nano Lett.*, 2016, **16**, 2268–2277.
50 M. W. Louie and A. T. Bell, *J. Am. Chem. Soc.*, 2013, **135**, 12329–12337.
51 C. Genovese, M. E. Schuster, E. K. Gibson, D. Gianolio, V. Posligua, R. Grau Crespo, G. Cibin, P. P. Wells, D. Garai, V. Solokha, S. Krick Calderon, J. J. Velasco Velez, C. Ampelli, S. Perathoner, G. Held, G. Centi and R. Arrigo, *Nat. Commun.*, 2018, **9**, 935.
52 X. Cheng, C. Lei, J. Yang, B. Yang, Z. Li, J. Lu, X. Zhang, L. Lei, y. hou and K. K. Ostrikov, *ChemElectroChem*, 2018, **5**, 1–8.
53 X. Zou, Q. Sun, Y. Zhang, G. D. Li, Y. Liu, Y. Wu, L. Yang and X. Zou, *Sci. Rep.*, 2018, **8**, 4478.

572815 Nickel Nanoparticles Embedded N-Doped Carbon Nanotubes As Biocompatible Electrocatalysis for CO_2 Conversion

Nickel Nanoparticles Embedded N-Doped Carbon Nanotubes as biocompatible electrocatalysis for CO_2 conversion

Siyuan Xiu, Gang Li, Yang Hou, Zhongjian Li*

College of Chemical and Biological Engineering

Zhejiang University

Hangzhou, China

Email: zdlizj@zju.edu.cn

Electricity generated from renewable energy, mainly solar and wind power, has drawbacks of discontinuity and fluctuation limited by natural conditions. It is an attractive strategy to store electrical energy in the form of chemical energy. Moreover, the capture and utilization of CO_2 contributes to eliminate the greenhouse effect. Water splitting-biosynthetic hybrid system for CO_2 conversion has been proved to be a promising technology for converting CO_2 into various chemical commodities and storing sustainable electricity into chemical energy simultaneously. The hybrid bio-inorganic system combines hydrogen evolution reaction (HER) catalysts with H_2-oxidizing autotrophic microorganisms. H_2 is produced at the cathodes with abiotic HER catalysts, then functions as an energy carrier and electron donor for hydrogen-oxidizing microorganisms. Intimate coupling of electrocatalysts and biosynthesis requires the catalysts possess both high catalytic performance and excellent biocompatibility, which is a bottleneck of the hybrid system. In this study, a complex of Ni nanoparticles embedded in N-doped carbon nanotubes (Ni@N-C) is synthesized as a HER electrocatalyst and is coupled with a hydrogen oxidizing autotroph, *Cupriavidus necator* H16, to convert CO_2 to poly-b-hydroxybutyrate(PHB). In Ni@N-C, the Ni nanoparticles are encapsulated in N-C nanotubes, which prevents bacteria from direct contact with Ni and inhibits Ni^{2+} leaching. As a result, Ni@N-C exhibits excellent biocompatibility and stability. The maximal PHB production of 384.30¡À6.53 mgL^{-1} with a maximum PHB production rate of 207.86¡À4.03 $g^{-1}m^{-2}L^{-1}day^{-1}$ were obtained, which is comparable to that of Pt/C. Our work demonstrates that this type of nanoconfined HER electrocatalysts reaches an excellent balance between electrocatalysis and biocompatibility through rational catalyst design.

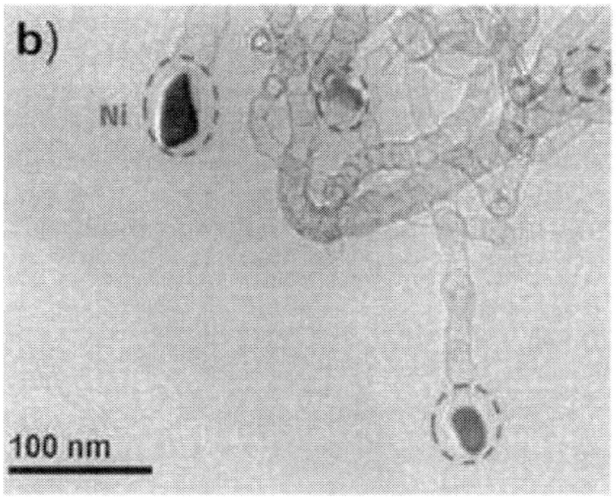

Figure 1 TEM of the Ni@N-C catalyst.

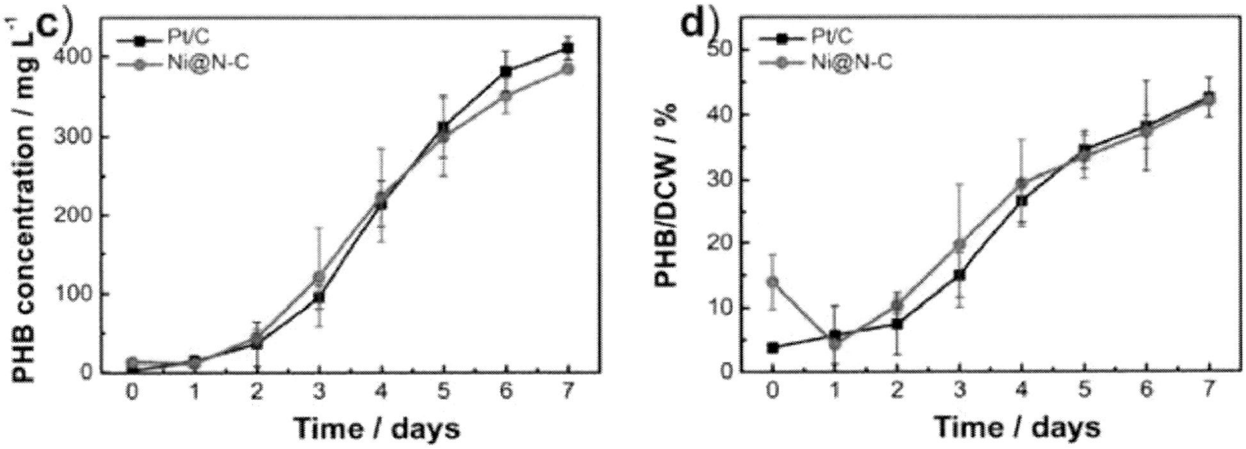

Figure 2 PHB c) PHB concentration, d) PHB content of dry cell weight

579052 Innovative Fluidized Bed Technologies in Coal to Chemical Processes

The increasing demand of chemicals including olefins and aromatics in China has triggered significant interests in developing coal via syngas and/or methanol to chemical processes. One of the examples is the methanol to olefins process (MTO) developed by the Dalian Institute of Chemical Physics (DICP), Chinese Academy of Sciences, which was successfully commercialized in 2010 in Shenhua Baotou's coal-to-olefins (CTO) plant (with an ethylene and propylene production capacity of 600 kt/a). In this DMTO process, fluidized bed reactor-regenerator configuration, akin to that in the fluid catalytic cracking (FCC) process, has been used. The DMTO reactor is a shallow fluidized bed with an aspect ratio of 0.3, representing a novel practice in industrial fluidized bed design. In this paper, we will discuss the progress of fluidized bed technologies in DICP, focusing on the applications in coal to chemical processes developed in recent years. We first report the fluidized bed technologies used in DMTO process, which include the shallow bed reactor, catalyst circulation rate control, and pre-coking technology. We then introduce the research and development (R&D) activities on fluidized bed reactor development and scale-up for methanol to propylene (DMTP) and methanol to aromatics processes. Beyond that, our strategy of innovative research for coal to chemical processes will also be highlighted.

576964 Methyl Formate Hydrogenation to Higher Hydrocarbons over Mn- and Ti-Oxide Promoted Cobalt Catalysts

Methyl formate hydrogenation to higher hydrocarbons over Mn- and Ti-oxide promoted cobalt catalysts

Authors: Andrew Raub, Motahare Athariboroujeny, Norbert Kruse

Methyl formate, the simplest of the carboxylic esters, has been considered since the early 1990s as a potential platform molecule for C1 chemistry[1]. Recently, we have established that it is possible to produce an array of chain-lengthened paraffins from methyl formate via a Fischer-Tropsch akin process utilizing cobalt-based catalysts at atmospheric pressure conditions. The fact that this molecule contains CO and H_2 in a 1:1 ratio makes it a potentially suitable feedstock for the preparation of hydrocarbon chains. In addition, methyl formate's low toxicity and relatively high boiling point (32 ^{o}C) make the molecule easy to transport and store. With further exploration, methyl formate hydrogenation may also provide a novel route for the synthesis of olefins and oxygen-functionalized hydrocarbons. In this work, the basic catalytic science of methyl formate hydrogenation over cobalt catalysts promoted by Mn- and Ti-oxides will be investigated. Manganese and titanium were selected due to their documented promoter effects observed for cobalt-based catalysts studied under conventional Fischer-Tropsch synthesis conditions[2, 3]. This paper will provide kinetic and mechanistic insight into the atmospheric pressure hydrogenation of methyl formate using transient response methods.

Chemical Transient Kinetics (CTK), as applied in the present studies, relies upon the abrupt switching between a non-reactive (H_2 + Ar) and a reactive gas mixture (H_2 + He + methyl formate) while maintaining constant molecular flow conditions at the inlet to a catalyst bed. The initial construction of the catalytically active surface, termed the "build-up", occurs when reactants are admitted to the gas phase over a freshly reduced catalyst surface. Likewise, the removal of reactants from the gas phase results in a "scavenging" of the surface as residual adsorbed species are hydrogenated and desorb. Observation of the build-up and scavenging behaviors for a catalytic system provides mechanistic insight unattainable from a steady-state kinetic analysis. Integration of calibrated surface flows allows for time-dependent carbon and oxygen coverages on the catalyst surface to be determined at different stages of the reaction.

During the build-up of the methyl formate hydrogenation reaction, using a Co-MnO_x catalyst under atmospheric pressure conditions at 220 ^{o}C, alkanes are observed to form in the sequence of their carbon numbers. With the exception of methane formation, which is the first product observed in the present study, chain-lengthened alkanes appear once methyl formate decomposes to produce CO in the gas phase. The time-correlation of chain lengthening initiation and CO appearance in the gas phase places a strong argument in favor of a CO insertion mechanism being in operation. Differing from earlier studies on methyl formate hydrogenation in our lab, in which only alkanes were observed during methyl formate hydrogenation over Co/MgO catalysts[4], we also observe olefins. Remarkably, the delay time for olefin formation is longer than that for alkanes. Methanol production is also observed, time-delayed relative to the appearance of gaseous CO. The formation of methanol from methyl formate under mild conditions using heterogeneous promoted cobalt catalysts has not been previously reported.

As for the reaction mechanism, we hypothesize that formyl groups (resulting from the dissociation of methyl formate) adsorbed on the catalyst surface undergo decarbonylation to yield CO, which subsequently inserts into the bonds of surface alkoxy or hydroxyl groups. The resulting carboxylate-type species may then be hydrogenated to produce alkanes/alkenes or turned into a (C+1)-alkoxy homologue subject to another CO insertion, and therefore, chain lengthening. Such a scenario would also be able to account for the formation of oxygenated products.

References

1. Lee, J. S.; Kim, J. C.; Kim, Y. G., Methyl Formate as a New Building Block in C_1 Chemistry. *Applied Catalysis* **1990,** *57* (1), 1-30.
2. Dinse, A.; Aigner, M.; Ulbrich, M.; Johnson, G. R.; Bell, A. T., Effects of Mn promotion on the activity and selectivity of Co/SiO2 for Fischer-Tropsch Synthesis. *Journal of Catalysis* **2012,** *288*, 104-114.
3. Morales, F.; de Smit, E.; de Groot, F. M. F.; Visser, T.; Weckhuysen, B. M., Effects of manganese oxide promoter on the CO and H-2 adsorption properties of titania-supported cobalt Fischer-Tropsch catalysts. *Journal of Catalysis* **2007,** *246* (1), 91-99.
4. Iablokov, V.; Kruse, N., Discovery of a Fischer-Tropsch Hybrid Reaction: Hydrogenation of Methylformate to Long-Chain Hydrocarbons with Anderson-Schulz-Flory Chain Length Distribution. *Chemcatchem* **2019,** *11* (4), 1200-1204.

570450 Gallate-Based Metal-Organic Framework As a Heterogeneous Catalyst for the Coupling of CO_2 and Epoxides

Carbon dioxide (CO_2) is the major greenhouse gas and its increasement in the atmosphere has evoked global environmental and societal concerns. The chemical transformation of CO_2 into value-added chemicals has attracted significant interest due to its great potential as a nontoxic, ubiquitous and sustainable C_1 resource. [1] The conversion of CO_2 into cyclic carbonates, which have wide applications as pharmaceutical, fine chemical intermediates, monomers for the production of polycarbonates, and aprotic polar solvents, is one of the most promising strategies because of its high atom efficiency. Although homogeneous catalysts have been industrialized for the production of cyclic carbonates, the processes still suffer from unviable catalyst recovery process and difficult product purification as well as extreme pressure and temperature conditions. [2] Hence, the design and preparation of highly efficient heterogeneous catalyst is of great significance to boost the cyclization of CO_2 and epoxides. Metal-organic frameworks (MOFs), which feature structural versatility, modularity, and amenability to be designed with specific functionality, have received much attention in catalysis, adsorption, and so on. [3-4] In view of their ordered porous structure, hydrophobic confined environment, and fruitful functionalizations, MOFs have been explored as one of the most appropriate materials for heterogeneous catalysis.

In the present work, we employed gallate-based metal-organic frameworks (M-gallate, M= Ni, Co, Mg, Mn) as a highly efficient and recyclable heterogeneous catalyst for CO_2 cycloaddition with epoxides in the presence of quaternary ammonium salts under solvent-free conditions. Experimental investigations revealed that Co-gallate shows superior catalytic activity to other M-gallate (M= Ni, Mg, Mn) and cobalt salts. This protocol is compatible with a plethora of terminal epoxides and consequently provides a practical alternative method for the fixation of industrial CO_2 into value-added cyclic carbonates.

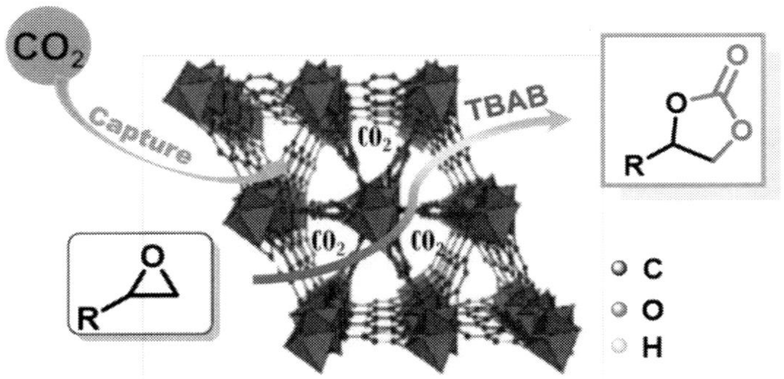

Scheme 1. M-gallate catalyzed CO_2 cycloaddition with epoxides

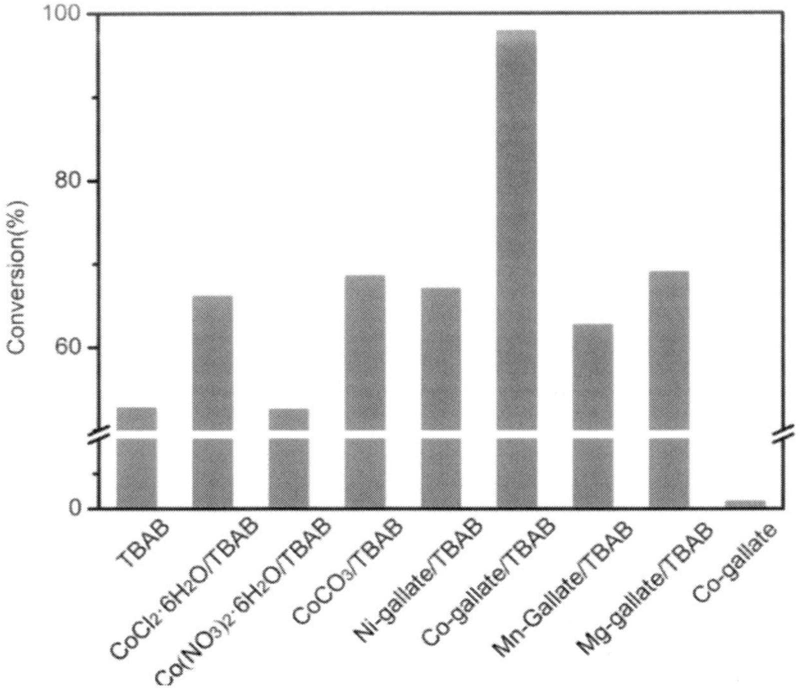

Figure 1. CO_2 cycloaddition with epoxide with different catalysts

References

[1] Song QW, Zhou ZH, He LN. Efficient, Selective and Sustainable Catalysis of Carbon Dioxide. *Green Chem*. **2017**; 19 (16): 3707-3728.

[2] Gao CY, A J, Tian HR, et al. An ultrastable zirconium-phosphonate framework as bifunctional catalyst for highly active CO_2 chemical transformation. *Chem. Commun.* **2017**, 53, 1293-1296.

[3] Kurisingal JF, Babu R, Kim S-H, et al. Microwave-induced synthesis of a bimetallic charge-transfer metal organic framework: a promising host for the chemical fixation of CO_2. *Catal. Sci. Technol*. **2018**; 8, 591-600.

[4] Babu R, Roshan, R, Kathalikkattil AC, et al. Rapid, Microwave-Assisted Synthesis of Cubic, Three-Dimensional, Highly Porous MOF-205 for Room Temperature CO_2 Fixation via Cyclic Carbonate Synthesis. *ACS Appl. Mater. Interfaces*. **2016**; 8, 33723-33731.

557675 Biological Upgrading of Methane for the Production of Chemicals and the Techno-Economic Analysis

The current crisis of global warming is primarily attributed to CO2 production from excessive use of fossil fuels during recent decades, and has increased demand for renewable biofuels tremendously. Lipids are drawing considerable attention in relation to the production potential of biodiesel on the basis of their nontoxic, sustainable, and energy efficient proprieties. However, the high cost of microbial lipid produced by oleaginous microorganisms mainly stems from the high cost of glucose, which is estimated to be about 80% of the total medium cost. Therefore, considerable efforts have been directed toward minimizing the carbon source cost and finding new alternative carbon sources. In this report, methane derived from biogas or natural gas was investigated for chemical production. After developing the culture modes and optimizing the culture conditions, both high titer and productivity were achieved in high cell density cultures of different microorganisms using various carbon sources.

In light of the relatively low price and the tremendous amount of methane and increasing demands of biochemicals and liquid transportation fuels, attention has begun to turn to novel biocatalyst for conversion of methane into bio-based products . A techno-economic analysis (TEA) was performed for an integrated process using fermentation process converting methane into chemicals such as lactic acid and isobutanol, which can be used as the pursuer for the production of chemicals and fuels. The entire process has been projected by using the Aspen plus software and the costs were calculated using modified excel spreadsheet. A sensitivity analysis was carried out based upon the identified key parameters, such as plant capacity, process technology (yields and rates), productivity, and raw material cost. This presentation also includes single-point sensitivity analyses and scenarios analyse results for several prospective cases. The presentation will also reveal that the major driving forces from operating cost and capital cost. The efforts of reducing the greenhouse emission from this process was estimated and compared with the traditional combustion of biogas and natural gas.

563733 In-Situ Hydrogen Sulfide Removal and Biogas Upgrading Via a One-Reactor System Integrating Solid-State Anaerobic Digestion of Food Waste and Bioelectrochemical Treatment

Food waste (FW) is estimated to account for 40% of the food produced in the United States, making it a critical and urgent environmental concern with a magnitude of 280-300 kg per year per capita. Among the approaches to valorize FW for biochemical/bioenergy production and to reduce its environmental footprint, solid-state anaerobic digestion (SSAD, total solids content = 15 wt%) offers the superiorities such as smaller reactors required, lower energy input for heating and stirring, easier disposal of digestate with lower moisture as biofertilizer or soil amendment, etc. However, with the higher organic contents accompany some shortcomings, typically including easy digester acidification and even failure caused by accumulation of excessive volatile fatty acids (VFAs), buildup of hydrogen sulfide (H_2S) in the biogas that is toxic to personnel and corrosive to facilities, etc. Conventional countermeasures such as extra addition of water or some bulk agents (e.g., yard wastes) reduce the capacity of FW treatment.

To fill in this knowledge gap, an innovative one-reactor system incorporating SSAD of FW and bioelectrochemical (BEC) treatment of leachate was studied. In the upper layer of this reactor, FW was hydrolyzed and VFAs were produced and washed down into the leachate. In the lower layer of reactor inserts the BEC unit. A pair of electrodes made of low carbon steel were submersed into the leachate and applied a voltage of 0.7 V, which secured the removal of most H_2S with relatively less material consumption of electrodes and increased the buffering capacity of leachate. A cyclic flow of leachate was created so that the BEC-treated leachate with a stronger buffering capacity was pumped back to upper layer for a more thorough mass transfer of SSAD. Moreover, the biogas profile was upgraded since the BEC treatment contributed to the reduction of some CO_2 to CH_4. Consequently, in-situ H_2S removal, biogas upgrading, and process stabilization were simultaneously achieved through this one-reactor design integrating SSAD of FW and BEC treatment of leachate.

564887 Rare Earth Elements Production from Geothermal Power Plant Using Magnetic Nanofluids

Extracting rare earth elements (REE) from geothermal brines is a very challenging problem due to the low concentrations of these elements and engineering challenges with traditional chemical separations methods involving packed sorbent beds or membranes that would impede large volumetric flow rates of geothermal fluids transitioning through the plant. We are demonstrating a simple and cost-effective magnetic nanofluid-based method for extracting rare earth metals from geothermal brines out from geothermal power plant. Core-shell magnetic nanoparticles are produced that contain a magnetic iron oxide core surrounded by a shell made of an Indium metal-organic framework (MOF) sorbent functionalized with chelating ligand diethylenetriamine (DETA) which is selective for the rare earth elements. The results of a study on the operating parameter matrix for the magnetic separator showed that it is possible to achieve a high (>99%) magnetic particle retention rate under appropriate combinations of magnetic power and liquid flow rate. It was found that the magnetic force may impact the integrity of the core-shell nanoparticle and lead to potential Fe leakage. The porous iron oxide core was replaced by a dense iron oxide core to greatly reduce the generation of Fe^{3+} in long term operation. The adsorbent life time was evaluated in both static and cycling manner by comparing the REE adsorption performance before and after extended hour (over 500h) experiment. In a case study, the internal return rate (IRR) of using the magnetic nanofluid extraction technology to separate Eu from Salton Sea geothermal brine was estimated to be over 18%, which indicates a promising economic feasibility.

560686 Adsorption Material for Thermal Energy Storage at Low Relative Humidity Operating Conditions

A wide range of energy storage technologies have been developed to increase the flexibility of the energy generation system [1]. Thermal energy storage (TES) provides an alternative method to store thermal energy generated from different types of energy sources (traditional or renewable) and correct for the mismatch between the energy supply and demand. Thermal energy storage through water vapor adsorption process has attracted increasing interest for space heating and cooling applications due to its safety and sustainability. Various kind of materials have been investigated in search of adsorbents with high energy density and stability.

In our previous study, silica gel has been identified as a promising adsorbent candidate for thermal energy storage applications at 80-120°C regeneration temperature range with relatively low relative humidity (RH) [2]. In this study, a commercial adsorbent powder was purchased and pelletized for adsorption experiments to examine the effects of the regeneration temperature and RH on the material behaviour, and its experimental energy density under different operating conditions. The preliminary experimental results showed that the energy density was three times as much as the energy density for the silica gel at low RH operating conditions. Modelling and simulation of adsorption process at different operating conditions will be discussed as well.

References:

[1] A. G. Olabi, "Renewable energy and energy storage systems," Energy, 136: 1–6 (2017).

[2] Y. Hua, A. Godin, and F. H. Tezel, "Water Vapor Adsorption in Silica Gel for Thermal Energy Storage Application," Adv. Mater. Lett., 10(2): 124–127 (2019).

576043 Superlocal Equilibrium in Low Temperature Plasma Reactors

Prominent methods of harvesting renewable energy involve the production of electricity for societal utilization. Compared to petroleum, which is a source of both heat and chemicals, renewable energy sources only provide work. In a sustainable society, electricity must be used to act upon material flows to increase the free energy of the compounds to make platform chemicals from which commodity products can be synthesized. Ideally, the synthesis of platform chemicals would involve removing carbon dioxide from the atmosphere. Low temperature plasma (LTP), which is a partially ionized gas in which extremely hot electrons with a temperature in the range from 10,000 to 100,000 K are immersed in a molecular background at 300 to 1000 K, comprises an open driven system capable of causing chemical transformations that proceed away from the local equilibrium state at the background temperature. Thus LTP is capable of increasing the specific free energy of mass flows. If one cannot use the idea of local equilibrium at the background temperature, then how can one predict the direction of a chemical reaction occurring in LTP? In this presentation, a proposed idea of superlocal equilibrium will be presented that can be used to make such predictions. The basic idea is that thermal equilibrium is local in both space and chemical species. Experimental evidence will be presented that in LTP, chemical reactions proceed towards a steady state that is entirely different from the local equilibrium state at the background temperature. The superlocal state depends on the background temperature, total pressure, electron temperature, plasma density, and relative amounts of different elements in the system; but is independent of species mol fractions at the inlet. Specifically, at low background temperature < 1000 K, it will be shown that the effluent from a plasma containing Ar + CO_2 approaches a steady state comprised of mostly CO + O_2, independent of whether CO_2 or CO+1/2O_2 were fed into the reactor. The experimental results were used to calculate the chemical potential of hot electrons present LTP at the conditions used in this work.

564659 Effects of Transition Metal Cation Additives on the Passivation of Lithium Metal Anode in Li-S Batteries

The effects of transition metal cation (Zn, Cu, Co, Ni, and Mn) additives on the passivation of lithium metal anode in Li-S batteries was investigated. The use of transition metal cations greatly improved the nature of SEI film, making it possible to suppressing further degradation of lithium metal. The specific discharge capacity with carbon nanofiber cathode after 100 cycles were found to be 873, 824, 767,751,718, and 694 mAh g^{-1} for Zn, Cu, Co, Ni, and Mn and no salt, respectively. The effects of transition cation additives on improving rate capability and Coulombic efficiency were observed in the following order: Zn > Cu > Co > Ni > Mn > no salt. SEM images indicated with Zn additive, a smoother surface and less accumulation of sulfur was observed compared to the one without additive. Furthermore, EQCM shows less accumulated mass on the anode surface for Zn additive as compared to Cu and no additive. The findings can be attributed to transition metal sulfides deposited on lithium surface cooperate with lithium sulfide and electrolyte decomposition product resulting in a smoother and more robust SEI film. The enhanced homogeneity of passivation layer greatly hinders the parasitic reactions between lithium metal and polysulfides as well as organic electrolyte, reducing the loss of active material and improving Coulombic efficiency. On the other hand, the uniform and mechanically strong SEI film introduces less undesired lithium plating and further accumulation on active sites, which significantly suppresses the dendritic formation due to improved surface morphology and chemistry. The results of this study provide a facile method of in-situ chemical formation of passivation layer protecting the lithium metal in batteries.

577355 Fibrous Conduction Networks Under Various Calendaring Regimes in Electrospun $LiFePO_4$/C Fibers

Advanced materials developments, in particular nano-structured materials, show large potential for battery improvements;but face challenges retaining those improvements when inserted into commercial production processes. In addition, battery material advances are often evaluated at low active mass loading per unit area leading to performance demonstrations at modest practical energy densities.

Calendaring is a typical process to compact battery electrodes to reduce thickness and increase electrode energy density. Calendaring reduces inter-particle distance within a cathode, improving short range electronic conduction and increases active material density; however, this improvement must be balanced against both reductions in void space (i.e. electrolyte penetration) and potential reductions in long range electronic conduction networks.

We use a fibrous carbon matrix as a conductive enhancement to Lithium Iron Phosphate/Carbon (LFP/C) cathode material, produced via electrospinning. This fibrous carbon offers a route to longer range conductive networks that maintain conductive paths under higher calendaring regimes. The electrospun composite LFP/C fibers are made up of thinly coated LFP crystalites within a fibrous carbon matrix. These electrospun fibers provide control over fiber diameter, crystallite size, and cystalite coating thickness for morphology/composition variation through a scalable process.

We report the resilience of conductive networks in electrospun fibrous LFP cathode material to various calendaring regimes, including the range of active mass loadings and void fractions. We review potential for increased practical energy density from electrospun fibrous LFP cathode materials.

579323 H_2 Generation Ability of Binary Transition Metal Perovskites via Thermochemical Water-Splitting Process

Perovskites with ABO_3 and $A_2B_2O_4$ structure due to their mixed valance state of the 3d transition metal sites with improved reduction and oxidation properties and large proportion of oxygen non-stoichiometric relating to high diffusivity of oxide ions, can be interesting materials to explore for H_2 production. In this study, we investigated $TiMnO_3$ and Co-doped $TiMnO_3$ powdered material for H_2 production *via* thermochemical water-splitting process. These perovskite materials were synthesized using sol-gel technique in which as-prepared gels were aged and calcined at 600°C in air. Phase purity and SSA of these perovskite materials before and after thermochemical water-splitting reaction was determined by XRD and BET SSA analyzer, respectively. As-calcined powdered perovskite material was loaded in the fully automated thermochemical Inconel reactor and water-splitting and regeneration steps reaction were performed isothermally at 1100°C. The crystal size and lattice strain were determined from the XRD analysis using Scherrer's equation and Williamson and Hall methods for the materials calcined at different temperatures. Charge storage ability of these perovskites was measured in terms of specific capacitance by performing cyclic voltammetry studies.

ChemPren: A new technology for the conversion of mixed municipal waste to basic chemicals and fuels

Dr. Piyush Sabharwall, Dr. Lynn Wendt, Ms. Roxanne Pinsky and Dr. Anne M. Gaffney
Idaho National Laboratory
Idaho Falls, ID 83404

Dr. Gennaro J Maffia, Amanda E. Peterman and Xufang (Katie) Liu
Manhattan College
Riverdale, NY 10471

Currently, over 90% of used plastics are not directly recycled and end up in U.S. landfills or incinerators. As more composite plastic materials are developed, this problem will only get worse. Technologies to up convert waste plastics to petrochemicals that can be used in the production of new plastics and materials, have been developed in the U.S., Japan, and Korea [1-4]. Catalytic conversion of waste plastics to Benzene, Toluene and Xylenes (BTX) was initially developed by ARCO chemical in the early to mid-1990s [1,2]. The BTX mixture is a versatile feedstock and can be used for the production of polyethylene terephthalate (PET), polycarbonates, polyurethanes, nylons, SBR, ABS and other polymers which are in turn used to produce beverage bottles, clothing, carpeting, automotive components and a broad range of other products. Intellectual property for improvements in the process and catalyst to up convert waste plastics to BTX was recently signed over to INL [5]. This technology capitalizes on INL owned intellectual property and advances the technical readiness level from the current level of 4 to a level of 7 with the project culminating in demonstrating at least 1000 hours of pilot plant operation.

References:

1. Jones, A.C., Gaffney, A.M., Leyshon, D.W., Cozzone, G.E., Sofranko, J.A., Process for reclaiming a thermoset Polymer, U.S. Patent No. 5,192,809, March 9, 1993.
2. Gaffney, A.M., Jones, A., Process for treating polyether polyols, U.S. Patent No. 5,462,971, October 31, 1995.
3. Nishino, J., Itoh, M., Fujiyoshi, H., Uemichi, Y, Catalytic degradation of plastic waste into petrochemicals using Ga-ZSM-5, Fuel 87 (2008)3681-3686.
4. Jung, S., Cho, M., Kang, B., Kim, J., Pyrolysis of a fraction of waste polypropylene and polyethylene for the recovery of BTX aromatics using a fluidized bed reactor, Fuel Proc. Tech. 91 (2010) 277-284.
5. Gaffney, A.M., Alloyed zeolite catalyst component, Method for making and catalytic application thereof, U.S. Patent Application 0215043 A1, August 23, 2013.

559476 Anaerobic Digestion of Food Wastes into Energy and Fertilizer

Anaerobic digestion (AD) has been evaluated as one of the most cost-efficient and environmentally friendly biotechnologies for renewable energy production, in the form of heat and power, organic waste management, and recovery of valuable nutrient rich digestate [1]. The latter has the potential to be used as fertilizer for agricultural utilization, landscaping, and soil amelioration since it contains essential nutrients such as nitrogen (N), phosphorus (P), and potassium (K) [2]. Digestate is produced from anaerobic degradation of organic wastes, mainly derived from food waste, crops and sewage sludge from wastewater treatment processes. However, the use of digestate has become a crucial challenge due to the limitation of cost and quality for the production of fertilizers as well as the application location. This study examined the optimization of an anaerobic digestion waste-to-energy system for food waste management and conversion of digestate to fertilizers. Following which, how the use of digestate for the cultivation of vegetables was investigated in comparison with chemical fertilizers. The objectives of this study are: 1. to optimize the operational parameters of a distributed anaerobic digestion waste-to-energy system for food waste management. 2. to study the efficiency of various fertilizers extracted from digestate and chemical fertilizers by cultivating green vegetables and comparing the major indexes for the growth of vegetables.

571999 Subcritical Water Upgrading of Coal Refuse with Food Waste

The goal of this research was to improve the fuel characteristics of coal refuse through co-hydrothermal carbonization (co-HTC) with food waste. Coal refuse, also known as coal waste and gob, are by-products resulted from coal mining. Coal processing plants reject 30-60% materials from the mined coal during the cleaning procedure and it contains 40-70% energy compared to coal. However, toxic elements (sulfur, heavy metals, and ash) present in coal refuse possess serious environmental threat and that is why it is not possible to combust coal refuse without prior treatment. Co-HTC of coal refuse along with food waste can be an efficient option to utilize this huge amount of available fuel resource. Co-HTC is a thermochemical process, where two feedstocks are treated at high temperature (180-280 °C) and corresponding vapor pressure (5-60 bars) for a short period of time. Coal refuse from local coal mine and food wastes from Ohio University Dining Hall were co-hydrothermally treated at three different temperatures (180, 230, and 280 °C) for 30 min. The acidic medium created by the hydrothermal carbonization of food waste catalyzed the upgrading of coal refuse. The elemental carbon (%) increased to a maximum value of 49.4 % for co-HTC solids from 18.3 % for raw coal refuses. Additionally, the ash percentage decreased from 66.4 % to 27.4 % for co-HTC solids. Moreover, the van-Krevelen diagram showed that the co-HTC solids have fuel characteristics similar to bituminous and sub-bituminous coals. The total sulfur concentration also decreased to less than 1.4 % from 9.9 %. The raw coal refuse samples and treated samples were further analyzed using the standard test method (ASTM D2492-02) and an induced coupled plasma optical emission spectroscopy (ICP-OES) to determine sulfate sulfur and pyritic sulfur concentration, respectively. The organic sulfur was calculated by subtracting the summation of sulfate and pyritic sulfur from the total sulfur concentration. The study also evaluated the distribution of sulfur among the co-HTC products. Later, an economic analysis was performed to check the economic feasibility of large scale hydrochar production. The economic study revealed that the production cost of co-HTC solids to be $39 per ton in the basis of a 110 MWe power plant.

556642 Polar Solvents Facilitate Efficient Microwave Pretreatment of Woody Biomass

Microwave pretreatment of woody biomass may be the key to unlocking an energy efficient process for recalcitrant waste stream for the production of renewable chemicals and fuels. Sugars derived from non-food sources such as woody biomass have the potential to be converted into a wide range of chemicals. However, woody biomass requires pretreatment to open the structure to access cellulose. Steam explosion, organosolv and sulphite pretreatment are currently used to remove lignin from wood [1]. However, these methods require substantial energy to be effective [1]. Development of an alternative, more efficient pretreatment processes for the release of sugars and other useful fractions from renewable biomass sources is necessary for a sustainable future.

Ionic liquids have been extensively researched for woody biomass pretreatment as they have proven to yield exceptional results via ionic conduction mechanism. However, limited research has been reported on the intentional exploitation of polar solvents, which increase the microwave absorption capacity via dipolar polarisation mechanism, for the pretreatment of recalcitrant biomass. Microwave absorption can vary significantly depending upon the ratio of effective dielectric loss factor (") to the dielectric constant (), of which the ratio is commonly known as the loss tangent (tan).

In this research, we investigated the use of solvents with high tan value to increase the efficiency of microwave-assisted solvolysis of woody biomass through an increase in the heating rates, and hence reduction of the energy consumption. To the best of our knowledge formaldehyde pretreatment of wood has not yet been reported under microwave irradiation. We studied the microwave irradiation of *Eucalyptus Regnans* and *Pinus Radiata* sawdust (~0.5 mm) in the presence of formalin solution containing 1.8, 3.7 and 10 wt% formaldehyde and water as the control at 180°C for 10 minutes.

It was found that 3.7 wt% formaldehyde (FA) solution is more efficient than water only for the pretreatment of wood sawdust at 180 °C and 10 min microwave exposure. The presence of formaldehyde in the solution accelerated the heating rate by 50% in the microwave. Furthermore, >75% of the hemicellulose sugars were released under mild microwave treatment conditions for both the wood species. Whereas in water-only pretreatment 32% and 55% of the hemicellulose was released from *E. regnans* and *P. radiata*, respectively.

A proposed mechanism for formaldehyde-mediated deconstruction of hemicellulose was developed. The lignin in the remaining sawdust from the FA pretreatment was analysed by 2D NMR, which detected the presence of stabilised lignin monomers. Whereas, with water-only pretreatment this does not occur. This stabilised platform may open-up a more advanced continuous process which could allow near theoretical lignin monomer yields through the subsequent hydorgenolysis.

576948 Comparative Leading Pretreatments for Improving Lignin Waste Valorization to Biobased Products

Biorefineries offer a potential solution to the current energy and environmental challenges facing our generation. However, there are still many technical barriers in developing a sustainable biorefinery. Due to the intrinsic recalcitrance of lignocellulosic biomass (LCB), the inherent value of LCB is highly dependent on deconstructing and valorizing lignin polymer in a biorefinery concept, and not just exploiting the more uniform carbohydrates. Lignin is the largest renewable aromatic carbon source and the second most abundant terrestrial polymer after cellulose. Valorizing lignin solid waste to value-added products shows the potential to make a sustainability of biorefineries. Despite these advantages and the significant progress, lignin waste remains as an untapped resource and the application of lignin valorization is relatively unexplored due to the heterogeneity of diverse lignin substrates. We hereby compared the leading pretreatments including sulfuric acid, steam explosion, liquid hot water, AFEX, and alkaline pretreatment for tuning lignin chemistry and improving lignin waste valorization to biobased products. Lignin characterization results showed that the lignin chemistry and reactivity obviously depended on the pretreatment employed. After the delignification of pretreated solid, the lignin stream was used as carbon source for polyhydroxyalkanoates (PHA) production by engineered *Pseudomonas putida* KT2440. Results showed that multilevel fractionation and utilization process can achieve high-value conversion of lignin for PHA with high yield. Based on the further characterization of lignin form different fractionation streams, the molecular mechanism has been proposed to unveil the improved processibility of lignin from multilevel utilization process. Overall, improving lignin waste valorization to biobased products by employing leading pretreatments provided a unique set of features to deliver a sustainable biorefinery process.

571984 Waste to Resource: Synthesis of Electrolyte and Electrode Material in Vanadium Redox Flow Battery from Leachate of Industrial Waste

Fly ash, a typical waste driven out of boilers in combustion plants, contains a large range of metal oxides, and leads to environmental consequence without proper treatment. In this study, we focus on a fly ash generated from combustion of oil refinery waste, which includes substantial amount of metal oxides (mainly vanadium oxide and nickel oxide). Instead of landfill, chemical leaching was applied to extract and separate metallic components in the fly ash. Because of the high concentration of vanadium ion inside, the leachate can be used as the electrolyte in vanadium redox flow battery (VRFB). Unlike vanadium-related product synthesized from leachate, this application avoids solid-liquid separation and minimizes the wastewater generation. Meanwhile, as the secondary metallic composition, nickel ions become a low-cost resource for synthesis of nickel-dope nanomaterial that can be applied in VRFB as electrode materials to enhance V^{3+}/V^{2+} and VO_2^{+}/VO^{2+} reactions.

567686 Synthesis of $LiNi_{0.6}Co_{0.2}Mn_{0.2}O_2$ Cathode Material By a Carbonate Co-Precipitation from Spent Lithium-Ion Batteries

Lei Xing (speaker)[1, 2], *Shuying Sun*[1,2], *Sen Lin*[1,2], *Ping Li*[1,2], *Jianguo Yu*[1,2]*

1. *National Engineering Research Center for Integrated Utilization of Salt Lake Resources, East China University of Science and Technology, Shanghai, China, 200237.*

2. *State Key laboratory of Chemical Engineering, East China University of Science and Technology, Shanghai, China, 200237.*

Speaker and corresponding authors: E-mail: xinglei8406@126.com, jgyu@ecust.edu.cn

Abstract:

The spent lithium ion batteries (LIBs) are classified as hazardous solid wastes because they contain toxic heavy metals and corrosive electrolytes. However, they contain various valuable metals such as copper, aluminum, magnesium, nickel, cobalt and lithium. Recycling of valuable metals from spent LIBs mitigate both resource shortage and environmental contamination problems [1]. So, the aim of this work is to present a promising approach for recycling high value-added metals from the cathode material of spent LIBs.

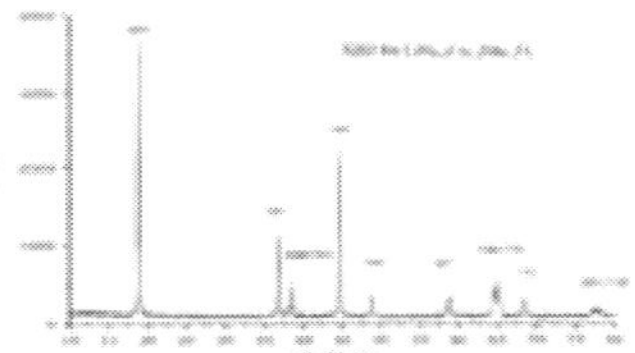

In this work, $LiNi_{0.6}Co_{0.2}Mn_{0.2}O_2$ cathode material is synthetized by a carbonate co-precipitation using the leaching liquid of spent LIBs as raw stocks. Firstly, the spent LIBs are manually dismantled to remove both the metal cases and plastic, and the cathodic material is separated from the Al foil by ultrasonic cleaning in the solvent 1-methyl-2-pyrrolidinone [2]. The valuable elements such as nickel, manganese and cobalt are obtained by leaching in sulfuric acid, and $NiSO_4$¡¤$6H_2O$, $CoSO_4$¡¤$7H_2O$ and $MnSO_4$¡¤H_2O are added into leaching solution to obtain the raw stocks for the $Ni_{0.6}Co_{0.2}Mn_{0.2}CO_3$ precursor through adjusting the molar radio of Ni: Co: Mn = 6: 2: 2. Secondly, the $Ni_{0.6}Co_{0.2}Mn_{0.2}CO_3$ precursor is synthesized by a carbonate co-precipitation method using the prepared raw stocks from leaching solution of spent LIBs. Thirdly, the $Ni_{0.6}Co_{0.2}Mn_{0.2}CO_3$ precursor is mixed with Li_2CO_3 at a Li/M (M=sum of Ni, Co and Mn) molar ratio 1.06, and the mixture is calcinated at 800-850¡æ under oxygen atmosphere to fabricate $LiNi_{0.6}Co_{0.2}Mn_{0.2}O_2$ cathode material. Finally, the advanced characterization methods (XRD, ICP-AES TG-DSC SEM-EDS) are used to demonstrate the synthetized $LiNi_{0.6}Co_{0.2}Mn_{0.2}O_2$ cathode material having a well-ordered layered structure and a low degree of cation mixing.

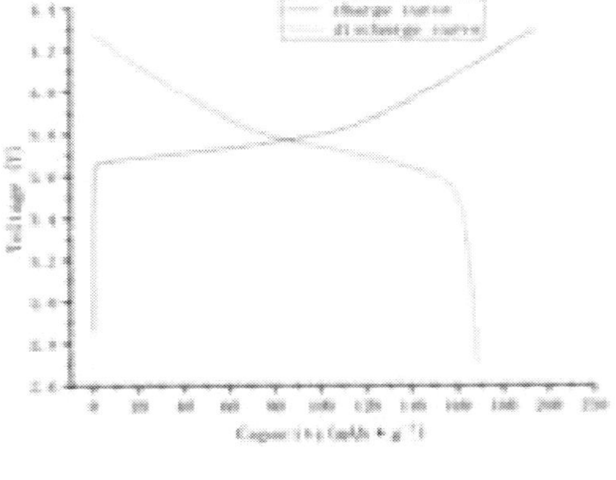

Furthermore, the electrochemical performance of the synthesized

$LiNi_{0.6}Co_{0.2}Mn_{0.2}O_2$ cathode material is tested in CR2025 coin cells. Electrochemical test methods include initial charge-discharge, continuous charge¨Cdischarge cycling, cyclic voltammetry and electrochemical impedance spectroscopy. It is found that the initial charge and discharge capacities of the $LiNi_{0.6}Co_{0.2}Mn_{0.2}O_2$ cathode material is 194.5mAh/g and 169.3mAh/g, and the initial coulombic efficiency is 87.05% when the initial charge-discharge curves are at a constant current density of 0.1C between 2.7V and 4.3V. The electrochemical performance of the prepared cathode material using the leaching liquid of spent LIBs is consistent with those of the cathode material synthetized from fresh materials. So, the proposed synthesis process has great potential for industrial-scale recycling from spent LIBs.

References:

[1] D.H.P. Kang, M. Chen, O.A.Ogunseitan, Potential environmental and human health impacts of rechargeable lithium batteries in electronic waste, Environ. Sci. Technol. 47 (2013) 5495-5503.

[2] L.H. He, S.Y. Sun, X.F. Song, J.G. Yu, Recovery of cathode materials and Al from spent lithium-ion batteries by ultrasonic cleaning , Waste Manag. 46 (2015) 523-528.

Keywords: lithium-ion batteries, cathode material, carbonate co-precipitation

557995 Sustainable Process for the Recycling of Polyethylene Phthalate

Polyethylene terephthalate (PET) is a thermoplastic polyester with excellent thermal and mechanical properties. PET is the third-most produced polymer (18% of world polymer production) and widely used in textile industry, X-ray films, food packaging particularly for water and soft-drink bottles.

Majority of plastic waste is going to land field. To recycle PET, chemical recycling has advantages over mechanical approaches in terms of better adaptability on heterogeneity of the solid wastes and no degradation of the product properties. In this paper, we will present a novel approach for chemical recycling of PET. This novel approach is featured by a simulated moving bed chromatography (SMBC) in the purification of Bis(2-Hydroxyethyl) terephthalate (BHET), which is the deploymerization product of PET. This novel approach can ensure a high purity of BHET that is suitable for reproducing chemical grade PET even with a low quality PET feedstock.

576903 Recycling Fe-P Slag As Energy Materials Via a Green Technics

Recycling waste is significant for the sustainable development of resource and environment. As one of the typical waste slags from phosphorus **industries, Fe-P generally has high melting point, plentiful iron and** phosphorus **elements and other impurities according to their raw materials and reaction technics. In order to recycle Fe-P slag as energy materials like $LiFePO_4$ and $FePO_4$, we have developed various routes including solid-state reaction and hydrothermal methods. The thermodynamics of different reaction processes are analyzed to obtain the possible routes with low energy consumption. Green technics with only water byproduct have been designed using the combined raw materials. $LiFePO_4$ and $FePO_4$ can be successfully obtained using $Fe_{1.5}P$ slag with ~1400°C melting point at no higher than 200°C. The purity and electrochemical performance of the as-synthesized samples are evaluated with various techniques. The conversion processes and key impact factors are discussed in detail. $FePO_4$ with good crystal structure and high purity can be obtained by adjusting the pH values of the product solution from the reaction of $Fe_{1.5}P$, H_3PO_4 and H_2O_2, which can meet the requirements for the production of $LiFePO_4$. On the other hand, $LiFePO_4$ can also be synthesized directly by using the raw materials of $Fe_{1.5}P$, H_3PO_4, H_2O_2 and LiOH. After being further coated with organic carbons using waste** sugarcane **slag, the as-obtained $LiFePO_4$ exhibits comparative electrochemical performance with that of current commercial samples. The discharge capacity can reach 150 mAh/g, and the capacity fade ratio is less than 1% after 100 cycles at 1.0C. The technics and equipment have also been proposed for mass production of $LiFePO_4$ and $FePO_4$ using Fe-P slag at low temperature. The green technics is interesting to recycle the waste with high melting point as function materials** through low-energy consumption way, **which is beneficial for the social sustainability with low environmental impact.**

569286 Recent Progress on Key Technical Barriers for Chemical Looping Combustion

Chemical Looping Combustion is an environmentally sustainable approach for converting abundant resources of fossil fuels. For the last five years, the U.S. Department of Energy's National Energy Technology (NETL) has been investigating this concept as an approach to mitigate greenhouse gas emissions in future power plant concepts. This presentation will review the progress in the chemical looping combustion technology development over the last five years with a primary focus on internal research and development efforts completed by NETL's Research and Innovation Center. The topic areas that will be discussed include the following:

- Results from plant simulations that identify key technology barriers for this technology
- Progress in oxygen carrier performance and durability which includes natural hematite, promoted hematite, and manufactured oxygen carriers
- Recommendations for advancing the technology readiness level for Chemical Looping Combustion in the future.

575572 Particle Properties in Techno-Economic Analysis of Chemical Looping Hydrogen Generation

Hydrogen (H_2) is considered as one of the most extensively used chemical intermediates in industry and a potentially attractive option for clean energy generation. However, CO_2 emissions from conventional H_2 production plants are significant, challenging the life cycle advantages of H_2 utilization. Thus, technologies are necessary to reduce or eliminate the carbon emissions associated with H_2 production while meeting the market demand in the present and future applications. Chemical looping hydrogen generation (CLHG) technology uses natural gas to generate H_2 with inherent CO_2 capture, which avoids contact between fuel and air (or oxygen) by utilizing a solid oxygen carrier to enable effective separation of O_2/N_2, CO_2 and H_2. CLHG can efficiently generate high purity H_2 with a high cold gas efficiency without the need for an amine-based CO_2 capture unit as required in conventional steam-methane reforming (SMR) or partial oxidation (POX) processes. In the CLHG process, iron-based metal oxide particles are circulated through a counter-current moving bed reducer reactor for full natural gas combustion to CO_2 and H_2O, a counter-current moving bed oxidizer reactor for reduction of steam by the reduced iron oxide to generate H_2, and a fluidized bed combustor reactor with riser for particle regeneration and transportation by air. After water removal, CO_2 can be separated from reducer flue gas and we can obtain high purity H_2 from oxidizer gas outlet. While particle is the key component of CLHG technology, this work focuses on the influence of particle properties, such as size, density, heat capacity, etc., on the overall CLHG process economics. Parametric studies of operating conditions were conducted to achieve auto-thermal operation of chemical looping reactor system and the pneumatic transportation of particles from combustor to reducer while varying particle properties. Specifically, this study developed the overall model of CLHG process with Aspen Plus simulation software. Several cases with representative particle properties and high process efficiency were configured. Energy and material balance information generated by the model was utilized to perform techno-economic analysis on the configured CLHG cases. The results indicate the requirement of particle properties to gain benefits of the CLHG system as compared to conventional H_2 generation process with CO_2 capture.

569092 High-Purity Hydrogen from Biogas with a Combined Reformer and Fixed-Bed Chemical Looping System

High-Purity Hydrogen from Biogas with a Combined Reformer and Fixed-Bed Chemical Looping System

The decentralized allocation of high-purity hydrogen plays a crucial role for a future hydrogen-based economy. Hydrogen will act as secondary energy carrier and temporary storage between fluctuating renewables as primary energy and the volatile demand from private and commercial consumers. The RESC process is proposed as a combination of a steam reformer and a fixed-bed chemical looping system within a single, compact reactor system for hydrogen generation by Hacker [1]. Depending on the syngas generation (i.e. reforming, gasification), different renewable and fossil feedstocks can be utilized. High pressure hydrogen generation was demonstrated in smaller lab systems reaching a hydrogen release pressure of up to 100 bar by Zacharias et al. [2].

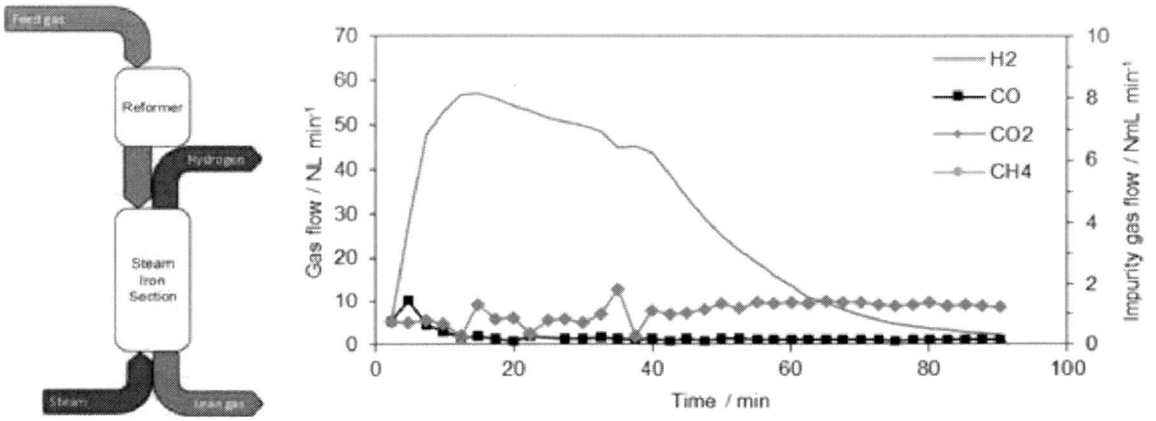

Figure 1: Schematic of RESC process (right) and exemplary hydrogen product gas in oxidation phase (left).

As part of the research activities, an experimental plant capable of hydrogen generation with an output power of 10 kW (LHV-based) from methane and 18 kg total inventory of oxygen carrier was presented by Bock et al. [3]. The system was capable of high purity hydrogen generation exceeding 99.999% from methane feedstock with carbon monoxide and carbon dioxide being the main contaminants. Ongoing research especially focuses on the utilization of biogenic resources as biogas or gasified biomass for hydrogen generation. Within the experimental research, the applicability of synthetic biogas mixtures was proved by Bock et al. [4]. Therefore, several biogas compositions were investigated regarding their suitability for hydrogen generation. The results confirm the high potential of biogas as potential source for high-purity hydrogen generation between 99.99%-99.999% purity as demanded by low temperature fuel cells (see Figure 1). Nevertheless, the application of synthetic biogas also led to increased carbon contamination compared to earlier experiments with methane as feedstock. The degree of utilization for hydrogen generation from synthetic biogas exceeded 40% within the experimental series, depending on the biogas composition. In addition, a variation of the process temperature was investigated in order to determine the influence on hydrogen purity as suggested by other research groups [5].

Long-term employment of the oxygen carrier also gives fundamental insights into the behavior for the proposed system. The applied material in the experimental plant featured a high resistance towards mechanical and thermal stress from cyclic reduction and oxidation in long-term operation throughout 70 cycles and more than four months of uninterrupted high-temperature exposure.

In addition, several trace gas compounds in biogas are prone to significantly affect the product gas purity. To investigate the impact of these compounds, the main impurities in common biogas and gasified biomass feedstocks were identified and tested. Their influence on carbon deposition and hence also on the product gas impurities in hydrogen generation phase was investigated in a fixed-bed system. Model compounds were defined for alkanes, cycloalkanes and alcohols and investigated in a fixed-bed system. Carbon depositions were identified especially from co-feeding alcohols and cyclic compounds at higher relative shares applied in reduction phase, although additional steam was applied to enable full conversion the compounds.

Acknowledgements

Financial support by the Klima- and Energiefonds through the Energy Research Program 2015 is gratefully acknowledged.

[1] V. Hacker, A novel process for stationary hydrogen production: The reformer sponge iron cycle (RESC), J. Power Sources. 118 (2003) 311–314. doi:10.1016/S0378-7753(03)00076-4.

[2] R. Zacharias, S. Visentin, S. Bock, V. Hacker, High-pressure hydrogen production with inherent sequestration of a pure carbon dioxide stream via fixed bed chemical looping, Int. J. Hydrogen Energy. (2019) 1–15. doi:10.1016/j.ijhydene.2019.01.257.

[3] S. Bock, R. Zacharias, V. Hacker, High purity hydrogen production with a 10kWthRESC prototype system, Energy Convers. Manag. 172 (2018) 418–427. doi:10.1016/j.enconman.2018.07.020.

[4] S. Bock, R. Zacharias, V. Hacker, Experimental study on pure hydrogen production from biogas in a combined fixed-bed reformer and chemical looping system, in review (2019).

[5] J. Lachén, P. Durán, J.A. Peña, J. Herguido, High purity hydrogen from coupled dry reforming and steam iron process with cobalt ferrites as oxygen carrier: Process improvement with the addition of NiAl2O4catalyst, Catal. Today. 296 (2017) 163–169. doi:10.1016/j.cattod.2017.04.046.

572284 Perovskite Oxygen Sorbents for "Low Temperature" Thermochemical Air Separation: Correlating Compositions with Redox Performance

As a potentially more efficient alternative to cryogenic air separation, thermochemical air separation via cyclic redox reactions of an oxide-based oxygen sorbent has been extensively investigated recently. Although a number of promising oxygen sorbents have been proposed and investigated, further improvements in sorbent performance through fundamental understanding of the relationships between oxygen sorbents' structural/compositional properties and redox performance are highly desirable.

In this study, we systematically investigated the effects of A and B site dopants on the oxygen uptake/release properties of $SrFeO_3$ family of perovskites as oxygen sorbents. Addition of dopants to the A site, namely Ba and La, or the B site, Cu or Mn, increased the oxygen capacity for temperature and pressure swings when compared to undoped $SrFeO_3$. Based on these results, the Mn sample underwent 1000 redox cycles where the sorbent was subjected to oxygen partial pressure swings between 20% O_2 and inert at 600 °C. The Mn doped sample proved to be stable in terms of oxygen storage capacity, with less than a 3% decrease in capacity. The redox kinetics were similar for the last few cycles when compared to the first few cycles. DFT calculations, in conjunction with the experimental results, indicated that oxygen vacancy formation energies are correlated to initial decomposition temperature for the samples investigated. However, the oxygen storage capacities of the doped samples did not show good correlation with the vacancy formation energies. It was further determined that the dynamic nature of the redox process makes it difficult to use a single vacancy formation energy as the descriptor for the oxygen storage capacities of the sorbents. Instead, a systematic approach was developed to correlate the oxygen storage capacities with the sorbents' compositional properties and vacancy formation energies. The findings from this study can potentially be used to develop improved sorbents for thermochemical air separation.

567742 MoVNbTeO for Chemical Looping - Oxidative Dehydrogenation of Ethane

Chemical looping (CL) is a concept that depends on conveying lattice oxygen, instead of gaseous oxygen, to a reaction by means of a metal oxide, termed, in this context, an oxygen carrier. This enables physical or temporal separation between the oxidation and reduction steps. Oxidation occurs when the oxygen carrier is contacted with air in the first reactor, while reduction takes place in the second reactor by reaction with a reducing agent. CL might be employed for the production of chemicals via selective oxidation, or oxidative dehydrogenation (ODH), reactions, which can render these reactions safer, more selective and more intensive. In addition, CL would result in reducing CO_2 and NO_x emissions. However, the key challenges to adopt CL for selective oxidations include quantifying the benefits and drawbacks of applying the concept, in addition to the development of a stable redox catalyst.

The present study investigates the potential of chemical looping ODH (CL-ODH) of ethane. A multimetallic catalyst, namely MoVNbTeO, was chosen as the primary catalyst for this study, given its high activity for ODH. The choice was confirmed by a wide catalyst screening, spanning other active metals and oxygen carriers. The activity of MoVNbTeO was optimised for CL by tuning preparation conditions, most importantly calcination regimes. XRD, TGA and SEM-EDS confirmed the improved crystal structure, high oxygen capacity and homogenous distribution of metals for the developed catalysts.

To further enhance ethylene yield, particulate oxygen carriers composed of bismuth oxides supported on ceria-zirconia were physically mixed with the catalyst. This class of oxygen carriers can selectively combust hydrogen in the presence of ethylene. The results showed that the presence of the oxygen carriers slightly improved conversion and CO_2 selectivity, without affecting overall ethylene yield.

Catalyst testing indicated that the developed catalyst can realise 50% ethylene yield, with 90% selectivity and 55% conversion at 450°C. This represents a significant improvement compared to CL-ODH catalysts described in the literature that can achieve similar yields only at temperatures as high as 850°C.

576625 Improving Fe_2O_3 Reactivity with CO and CH_4 in Chemical Looping Systems By Low Percentage 1st-Row Transition Metal Dopants: A Theoretical and Experimental Work

Chemical looping (CL) technology, which includes chemical looping combustion (CLC) and gasification (CLG), is one of the most promising platform in the field of clean energy due to its capability in simplified inherent CO_2 capture and value-added chemical production. One crucial factor in dictating the success of a CL process is the reactivity of oxygen carriers. In the CL technology developed at The Ohio State University (OSU), iron oxide is employed as the active oxygen carrier due to its abundance and low cost. In this work, we focus on improving the reactivity of Fe_2O_3 with CO and CH_4 by using low percentage (1%) 1st-row transition metal (TM) dopants. A systematic dopant screening strategy is developed via the combination of *ab initio* density functional theory (DFT) simulations and experimental validations conducted in TPR (temperature programmed reduction). Reaction pathways of CO and CH_4 oxidation on the most thermodynamically stable Fe_2O_3 (0001) surface with and without dopants and defects were investigated, and the rate-limiting steps (RLS) were identified. Descriptors that correlate the energy barrier of the RLS with different dopants were found. Our results demonstrate that copper is the most efficient dopant among 1st-row TMs as it can lower the reduction initiation temperature with CO and CH_4 by 50 and 150 °C, respectively. Such a decrease in operating temperature can expand the operational matrix for the autothermal operation, and facilitate the development of CL processes.

564782 CO_2 Conversion to 3D Graphene for Clean-Energy Application

Abstract: In the recent year, we discovered several reactions for the synthesis of 3D graphene [1-5]. In the presentation, I will highlight recent findings for CO_2 conversion to 3D graphene in my group: (1) CO_2 can react with alkaline metals (Li, Na, or K) to form various types of 3D graphene; (2) the synthesized 3D graphene materials possesses unique properties (such as high electrical conductivity and large accessible surface area); and (3) the 3D graphene materials exhibited excellent performance as electrodes for solar cells, supercapacitors, and batteries.

References:

1. H. Wang, K. Sun, F. Tao, D. J. Stacchiola, Y. H. Hu, "3D Honeycomb-Like Structured Graphene and Its High Efficiency as Counter Electrode Catalysts for Dye-sensitized Solar Cell", *Angew. Chem. Int. Ed.* 52, 9210(2013).
2. W. Wei, K. Sun, L. Cheng, A. J. Pak, E. Paek, G. Hwang, Y. H. Hu, "The Bright Future for Electrode Materials of Energy Devices: Highly Conductive Porous Na-Embedded Carbon", *Nano Lett.* 16, 8029(2016).
3. L. Chang, D. J. Stacchiola, Y. H. Hu, "An Ideal Electrode Material, 3D Surface-Microporous Graphene for Supercapacitors with Ultrahigh Areal Capacitance", *ACS Appl. Mater. Interface* 9, 24655(2017).
4. W. Wei, B. Hu, F. Jin, Z. Jing, Y. Li, A. A. G. Blanco, D. J. Stacchiola, Y. H. Hu, "Potassium-Chemical Synthesis of 3D Graphene from CO_2 and Its Excellent Performance for HTM-free Perovskite Solar Cells", *J. Mater. Chem. A*, 5, 7749 (2017).
5. L. Chang, K. Sun, Y. H. Hu, "New Chemistry for New Material: Highly Dense Meso-Porous Carbon Electrode for Electric Double-layer Capacitors with Ultrahigh Areal, Gravimetric, and Volumetric Capacitances", *ACS Appl. Mater. Interface*, 10, 33162(2018).

570669 Waste Plastics to Energy Storing Carbon Anodes through Green Processing

For the methodical recycling of abiquitous plastic waste, greener manufacturing or chemical processes are required for the environmental and societal benefits[1]. Sustainable engineering is the process of designing a system(s) in which they use energy and resources sustainably; ultimately generating an ability of future generations to meet their own needs. Environmentally ***harmful plastic waste*** is a major concern currently throughout the world.[2] Though plastics improve our lives in countless ways, they also pose a **serious** threat to our environment as plastic waste. However, the major challenges faced in recycling are (i) plastic waste collection, segregation, cleaning with associated cost, (ii) chemically different polymers extrusion yields low quality product and (iii) restrictions and imparted duties by other countries on import of recyclable plastic waste.

During last decade, Pol's research team innovated and researched on an environmentally beneficial thermochemical upcycling[3] processes with possible important applications for the obtained products contributing to the sustainable engineering field. Developed one-step, low-energy, solvent-less process is called **"upcycling"** because it produces products[4] having *greater value* than the original plastics. Developed transformative processing method could **permanently** reduce pollution at its source, providing **greener reaction conditions.** Established **greener** processing system for the conversion of unsorted plastic waste into structurally tunable carbon microspheres[5] and carbon nanotubes[6] that have important tribological[7] and battery applications[5,8] will be discussed. Moreover, recent **invention** lead by PI transforms polymer and starch waste, including used packing peanuts, into a multi-functional carbon sheets powder via an **environmentally friendly**, inexpensive, and scalable production[9] route. The manufactured carbon sheet particles demonstrate a wide-range of potential applications including: **electrodes** for energy storage, toners for printers, activated carbons for **storing CO_2**, conductive additives for tires, lubrication additives, and many more. This industrially viable, facile **green process** utilizes trashed packing peanut feedstock that is heat treated to 600 °C under an inert atmosphere in a single-step attaining reproducible product as promising lithium[9] and sodium[10] ion battery anode. In summary, this talk is centered on the originated **green processing,** which applies economic and technologically feasible parameters to engineer a product decreasing pollution at its source, minimize hazardous exposures, and reduce energy consumption protecting human health.

References

1. United Nations Environment Assembly. *Towards a pollution-free planet*. Report No. UNEP/EA.3/L.19 (United Nations, 2017).
2. United Nations. *Transforming our World: The 2030 Agenda for Sustainable Development*. Report No. A/RES/70/1 (United Nations, 2015).
3. V. G. Pol, Upcycling: Converting Waste Plastics into Paramagnetic, Conducting, Solid, Pure Carbon Microspheres. *Environ. Sci. Technol.*, **2010**, *44*, 4753–4759.
4. V. G. Pol, J. Wen, K–C. Lau, S. Callear, D.T. Bowron, C–K. Lin, S. A. Deshmukh, S. Sankaranarayanan, L. A. Curtiss, W. I. F. David, D. J. Miller, M. M. Thackeray, Probing the Evolution and Morphology of Hard Carbon Spheres. *Carbon*, **2014**, *68*, 104-111
5. V. G. Pol, M. M. Thackeray, Spherical Carbon Particles and Carbon Nanotubes Prepared by Autogenic Reactions: Evaluation as Anodes in Lithium Electrochemical Cells, *Energy & Environ. Sci.*, **2011**, 4, 1904–1912
6. V. G. Pol, P. Thiyagarajan, "Remediating Plastic Waste into Carbon Nanotubes," *J. Environ. Monitoring*, 2010, 12, 455–459.
7. K. K. Mistry, V. G. Pol, M. M. Thackeray, J. Wen, D. J. Miller, A. Erdemir, Synthesis and Tribology of Micro-Carbon Sphere Additives for Enhanced Lubrication, *Tribology Transactions*, **2015**, 58, 3.
8. Pol, V. G.; Lee, E.; Zhou, D.; Dogan, F.; Calderon-Moreno, J. M.; Jonhson C. S. Spherical Carbon

as a New High-Rate Anode for Sodium-ion Batteries. *Electrochim. Acta*, **2014**, *127*, 61-67.
9. Etacheri, C. N. Hong, V. G. Pol, Upcycling of Packing-Peanuts into Carbon Microsheet Anodes for Lithium-Ion Batteries. *Environ. Sci. Technol.* **2015**, *49*, 11191.
10. J. Tang, J. Barker, V. G. Pol, "Sodium-ion Battery Anode Comprising Carbon Sheets: Stable Cycling in Half and Full-Pouch Cell Configuration", *Energy Technology*, 2018, 6, 213-220.

560332 Molecular Simulation and Experimental Study of CH4 Adsorption in Surfactant Solution

Molecular Simulation and Experimental Study of CH4 Adsorption in Surfactant solution
Yihan Xue
Beijing Key Laboratory of Urban Oil and Gas Distribution Technology, China University of Petroleum (Beijing), Beijing, China
Email address:
xyh3610@126.com(Yihan Xue)
Abstract: The research on the storage and transportation technology of green methane is a hot spot nowadays. In order to overcome the disadvantages of LNG and CNG, the effect and method of methane adsorption in surfactant aqueous solutions are studied. In order to compare the effect of the methane adsorption in surfactant aqueous solutions, five surfactant aqueous solutions of sodium dodecyl sulfate (SDS), polyoxyethylene sorbitan monooleate (Tween 80), octylphenol ethoxylate (TX-100), dodecyl dimethyl betaine (BS-12) and castor oil polyoxyethylene ether (EL-20) at the concentration range of 25 g/L-200 g/L are selected for the study. Molecular simulation and experimental research are adoped in the research process. The adsorption energy and adsorption amount in five surfactant systems with different concentrations is numerically simulated at normal pressure or low pressure with normal temperature, at the same time, experimental study is performed. The numerical simulation results agree well with the experimental results. The results show that SDS has the best adsorption effect on methane at the same concentration. According to the research results, suggestions are made for the industrial application of methane adsorption in surfactant aqueous solutions.
Keywords: Methane storage and transportation, Surfactant aqueous solution, Adsorption energy, Adsorption amount

562653 A Comparison of Two Bioconversion Processes for the Production of Commercially Important C_4 Streams

Four carbon compounds like 1,4-butanediol (1,4-BDO), succinic acid (SA) have been identified as platform chemicals for producing polymers from renewable resources. The successful commercialization of 1,3-propanediol by Dupont Tate & Lyle Bioproducts and 1,4-BDO by Genomatica Inc. has established the commercial viability of green chemicals. In this talk we present the techno-economic analyses to produce four-carbon compounds (C_4s) like 1,4-BDO, gamma butyrolactone (GBL), and tetrahydrofuran (THF) from dextrose through two separate routes - (i) single step bioconversion and (ii) hydrogenation of bio-based succinic acid (SA). In the single step bioconversion process for production of 1,4-BDO, capital cost of the bioconversion section was found to contribute to around 58 % of the installed equipment cost. Hydraulic load from bioconversion section and membrane filtration was found to contribute the most to operating cost. The process was found to be economically feasible with the minimum sales price (MSP) of 1,4-BDO of $1.82/kg as compared to a market price of $2.5/kg. A mechanical vapor recompression scheme used in the evaporator section was found to benefit the process, increasing energy efficiency and cost effectiveness. In comparison, the SA hydrogenation process, called the C_4s refinery, was not economically competitive given the current state of technological progress. The C_4s refinery produces 1,4-BDO, GBL, THF and ammonium sulfate (AS) fertilizer. The capital costs of the C_4s refinery were found to be 1.8 times higher while the operating costs were found to be 2.5 times higher. An important parameter is the SA fermentation titer. Increase in SA titer values to the maximum values reported in literature were not able to reduce the price of C_4s to below the market price, demonstrating the need for progress in this area. It was concluded that for an extended process involving hydrogenation of SA, additional increases in SA titer values and productivity by genetically modifying the relevant organisms will be required to achieve viability especially in comparison to the single step fermentation process for 1,4-BDO. A secondary reason for unfavorable economics was found to be the dependence of the C_4s refinery on thermal separations which contribute to significant operating and capital costs, technological progress in separation processes such as membranes for organic water separations is therefore necessary. Overall, given the published state of technology, the direct fermentation route produces C_4s (1,4-BDO) at $ 1.82/Kg and is economically viable while the C_4s refinery produces C_4s at $ 3.67/Kg and is not competitive with current production routes.

572716 An Integrated Tool for Rapid Assessment of Chemical Manufacturing Process Emissions, Treatment and Sustainability Performance

The design and assessment of chemical manufacturing processes have been aided by a variety of tools that enable the comprehensive analysis and performance testing of both the steady and transient responses of processes with different levels of complexity. However, comprehensive tools that encompass the evaluation of process emissions, treatment simulation, and sustainability assessment of chemical plant performance are yet to be presented.

In this contribution, the tools developed by the co-authors for simulating Pollution Control Units (PCUs) [1] and performing sustainability assessment with GREENSCOPE [2] are combined for the first time. The PCUs allow for the simulation of several unit operations associated with emission treatment of gas, liquid and solid streams [1], while GREENSCOPE enables sustainability analysis considering four categories: Environmental, Mass, Energy, and Economic Efficiency. Each category has several unique metrics called indicators which provide insight into the sustainability performance of any chemical process. The new integrated tool thus allows for the analysis of scenarios including the impact of current and future models of unit operations on process economics, environmental sustainability, and emissions.

This work therefore presents a unified software framework that provides the opportunity to simulate and perform emission control analysis, pollution control unit sizing and sustainability assessment that could be used to support risk evaluation for different chemical manufacturing scenarios. The developed tool is illustrated via appropriate sections of a biomass/coal gasification process for syngas production with the end goal of acetic acid manufacturing.

References

1. Li S, Feliachi Y, Agbleze S, Ruiz-Mercado GJ ,Smith RL, Meyer DE, Gonzalez MA, Lima FV. A process systems framework for rapid generation of life cycle inventories for pollution control and sustainability evaluation. Clean Technol. Envir., 20(7):1543-1561, 2018.
2. Ruiz-Mercado GJ, Gonzalez MA, Smith RL. Sustainability indicators for chemical processes: III. biodiesel case study. Ind. Eng. Chem. Res. 2013;52(20):6747–60.

577369 Exploring the Potential of Chemical Looping Systems to Provide Alternative H_2 Production Systems Solutions Using Process Simulations and Techno-Economic Analysis

The iron-based moving bed based chemical looping technology developed at the Ohio State University (OSU) is a promising candidate to yield cost benefits in natural gas utilization for a carbon-constrained scenario. OSU has developed four commercially relevant, distinct approaches for high efficiency H_2 production from natural gas[1,2,3]. The first approach focuses on a three-reactor scheme that produces CO_2 and H_2 in separate reactors, providing process intensification benefits[1]. The second approach utilizes a membrane based separation scheme to surpass the thermodynamic limits associated with the H_2 production system in the first approach[2]. The third approach retrofits the methane to syngas based co-current reaction system in a steam methane reformer and considers co-feed of steam and CO_2 for high efficiency H_2 production[3]. A fourth approach focuses on modifying the first three reactor approach and producing syngas from the reducer in addition to the Hydrogen from the oxidizer. This presentation will initially examine, how the application of these four chemical looping systems can lead to process intensification and integration opportunities. The thermodynamic performance will be further quantified by a cold-gas efficiency and an energy efficiency comparison between the three approaches and a baseline Steam Methane Reforming case with 90% Carbon capture. Relevant experiments that include high-pressure kinetic data and sub-pilot scale experimental demonstration will be used as an input for a base techno-economic analysis model for the four approaches. The integrated process and techno-economic analysis conducted with guidance from industrial partners useful and surprising insights into how chemical looping Hydrogen production approaches provide solutions under different adoption and carbon-capture scenarios.

References

1. Kathe, M. V., Empfield, A., Na, J., Blair, E., & Fan, L. S. (2016). Hydrogen production from natural gas using an iron-based chemical looping technology: thermodynamic simulations and process system analysis. *Applied Energy*, *165*, 183-201.
2. Nadgouda, S. G., Kathe, M. V., & Fan, L. S. (2017). Cold gas efficiency enhancement in a chemical looping combustion system using staged H 2 separation approach. *International Journal of Hydrogen Energy*,Â *42*(8), 4751-4763.
3. Kathe, M., Empfield, A., Sandvik, P., Fryer, C., Zhang, Y., Blair, E., & Fan, L. S. (2017). Utilization of CO_2 as a partial substitute for methane feedstock in chemical looping methaneâ €“steam redox processes for syngas production. *Energy & Environmental Science*.

570829 Economic Analysis of Battery (NMC) Cathode Material Production in Flame Spray Process with Sustainable Solvents

The cost of cathode materials contributes approximately 32.7% to the cell construction cost of lithium-ion batteries, which has a significant impact on the price of battery packs. The synthetic routes of cathode material ($LiNi_{1/3}Mn_{1/3}Co_{1/3}O_2$ (NMC333)) based on the solid-state method or co-precipitation method involve complex mixing, drying, and sintering processes, resulting in a material cost too high to reach the U.S. Department of Energy battery price target of $125/kWh. To reduce the cathode material manufacturing cost, our lab has developed a flame spray process that utilize sustainable solvent glycerol as solvent to manufacture the cathode material. The economic analysis of the new green chemical process was studied based on discounted cash flow method. The major economic indicator used in this study is the minimum cathode material selling price (MCSP), whereas the co-precipitation method for NMC333 production was selected as the reference for economic analysis. Results show that compared with the co-precipitation method, the new flame spray process has the potential to reduce the operation cost by 31% with comparable capital investment and material cost. Moreover, MCSP of the new process is $19.1/kg, which is 83% of that of the co-precipitation method. When the flame spray process is integrated with the sintering process inline, the MCSP is as low as 15.6 $/kg, which demonstrates that the new process is an attractive manufacturing process for cathode material production. Furthermore, sensitivity analysis and uncertainty analysis were used to examine the effects of plant scale, raw material price, and analysis parameters on the MCSP and evaluate the certainty interval of the MCSP of the flame spray process.

574815 Soybean Oil As a Bio-Solvent to Sweeten Sour Natural Gas

Soybean oils are readily available bioresources composed of unsaturated fatty acids capable of binding sulfur compounds. Predictive models offer an opportunity to develop cost-effective methods using bio-based solvents to sweeten natural gas. While increases in fracking have increased availability of natural gas – a fuel that when combusted produces less greenhouse gases than coal or gasoline – natural gas is rich in hydrogen sulfide, rendering it sour gas. Hydrogen sulfide is corrosive to processing equipment and poses risks to human health and the environment. Predictive models such as COSMO-RS, the conductor-like screening model for real solvents, can be used to examine interactions at a molecular level, screen solvents, and choose process parameters. COSMO-RS was used to simulate the partitioning of hydrogen sulfide between soybean oil, high oleic soybean oil, free fatty acids, and gas phases. Predicted partition coefficient (K) values ranged from 0.16 to 0.70 for soybean and high oleic soybean oils at temperatures from 10 - 100°C; K values between 0 – 1 are indicative of hydrogen sulfide's preference for the liquid phase over the gas phase. Experimental results demonstrate the feasibility of removing H_2S from gas using soybean oils and are compared to the predicted partitioning. This study provides proof-of-concept for using high oleic soybean oil as a bio-based solvent as a new method for cleaning sour gas.

574828 Reducing Energy, Solvent, and Resin Demand of Sorption-Based Processes for Carboxylic Acid Recovery

Current methods such as ion exchange and distillation to separate bio-based carboxylic acids are material and energy intensive, typically generating waste streams and requiring energy intensive distillation steps to recover a pure product. Sorption-based approaches increase the sustainability of carboxylic acid separations by reducing downstream waste handling, however, current sorption-based technologies typically require large volumes of elution solvent and still result in a dilute carboxylic acid stream. Thus, to obtain a concentrated product, excess water and solvent must be removed from the elution stream via evaporation or distillation which adds a substantial operational cost to the process. This presentation will discuss recent work towards reducing solvent, adsorbent requirements and reducing downstream distillation energy demands in these processes. Specifically, four commercially available weak-base resins were screened by determining their adsorption capacity for butyric acid, a fermentable acid targeted for biofuels. Then, to recover the adsorbed acid, non-aqueous solvents were tested and their regeneration constant were determined. Findings indicate that specific properties of the resins and eluents correlate with the energy, resin, and solvent demands of the system. These results provide a general framework for optimizing sorption-based processes for recovering dilute bioproducts.

572591 Storage of Excess Wind Energy and Utilization of Waste Flue Gas CO_2 through Sustainable Methanol Production

As a larger production of wind energy and other types of renewable energies are predicted in the future, utilizing the excess energy from these sustainable energy sources can increase the energy efficiency of the wind farm by making good use of all the electricity produced. Moreover, CO2 emissions from industry, that contribute to climate change, are also an increasing problem that requires a solution. In this study, we present a novel concept, which aims at solving these two problems by integrating excess energy from wind farms and CO2 from flue gas into an innovative production process using methanol as an energy storage/carrier. The excess wind energy is used to produce hydrogen through water electrolysis and CO2 is captured from flue gas, e.g. from the cement industry. Compared with the traditional methanol production which heavily relies on natural gas as its source of hydrogen, carbon, and energy, this process has a smaller environmental footprint. The novelty of our concept relies on shifting the focus from operating a methanol production plant towards valorizing the excess renewable energy, which is wasted otherwise and considering methanol as a value-added product (as opposed to Szima et al. 2018). Therefore, the plant capacity of 30 000 metric tons per year is scaled based on the quantity of the excess wind energy of the largest wind farm in Denmark. This capacity is significantly lower than those of traditional methanol production plants, which range up to 1.9 million metric tons per year (Ferrostaal, 2015). This adds a constraint to improve the process integration efficiency of the proposed concept. Hence, a techno-economic analysis is performed to investigate the overall feasibility of the proposed concept by identifying key design decisions and essential cost drivers. As results obtained show high sensitivity to both electricity and methanol prices, an uncertainty analysis was conducted in order to determine their effects on the profitability metrics of the proposed novel process. A sustainability analysis is complementarily carried out to emphasize the added environmental benefits of the proposed concept in comparison to the traditional methanol production process.

Reference:

Szima, Szabolcs, and Calin-Cristian Cormos. "Improving methanol synthesis from carbon-free H2 and captured CO2: A techno-economic and environmental evaluation." *Journal of CO2 Utilization* 24 (2018): 555-563.

Ferrostaal. “M5000 Methanol Plant” from *www.ferrostaal.com/en/top-references-ferrostaal/m5000-methanol-plant/* (2015), visited 12.04.2019.

575320 Towards a Novel Energy Financial Security: The Texas A & M Energy Spot Price Index

The available financial instruments providing exposure for the investors to the energy markets are limited to equity based stocks (i.e. S&P 500 Energy) and futures on individual energy commodities (i.e. WTI Crude Oil, Henry Hub Natural Gas). Furthermore, these options are predominantly associated with the oil and gas sector, with almost zero emphasis placed on the renewable energy. Given the lack of a financial security to effectively capture the entire U.S. energy landscape and represent the average cost of energy, we introduce the Texas A&M Energy Spot Price Index (ESPI) as a novel framework to fulfill these needs.

The energy landscape of the U.S. is carefully depicted to determine the products that are directed to the end-use sectors of the U.S. economy [1-5]. The total energy demand of these products, together with their spot prices in the market, serve as the backbone of the Energy Spot Price Index. Since the weights of the Energy Spot Price Index constituents have a lag between two and three months, we introduce a rolling horizon based parameter estimation model to determine the current weights. The spot prices of the products are taken from Bloomberg when available. For products whose spot prices are not available, we introduce and examine three different approaches [6-8]. The methodology is illustrated on 120 months to show the capability of accurately determining the value of the Energy Spot Price Index. The proposed Energy Spot Price Index is the first of its kind that precisely quantifies the average market price of energy. As such, it can be utilized as an efficient tool for all investors who look for exposure over the average market price of energy, trading over the whole energy market, hedging their investments, as well as a benchmark for measuring the overall cost of energy. Therefore, the Energy Spot Price Index can be used as a tradable financial security (e.g. ETF) that is updated daily providing an alternative for the investors who look to diversify their portfolios.

References:

- Elia, J. A.; Baliban, R. C.; Xiao, X.; Floudas, C. A. Optimal energy supply network determination and life cycle analysis for hybrid coal, biomass and natural gas to liquid (CBGTL) plants using carbon-based hydrogen production. Computers and Chemical Engineering 2011, 35, 1399-1430.
- Elia, J. A.; Baliban, R. C.; Floudas, C. A. Nationwide energy supply chain analysis for hybrid feedstock processes with significant CO2 emissions reduction. AIChE Journal 2012, 58 (7), 2142-2154.
- Elia, J. A.; Floudas, C. A. Energy Supply Chain Optimization of Hybrid Feedstock Processes: A Review. Annual Review of Chemical and Biomolecular Engineering 2014, 5 (1), 147-179.
- Onel, O.; Niziolek, A. M.; Floudas, C. A. Integrated biomass and fossil fuel systems towards the production of fuels and chemicals: state of the art approaches and future challenges. Current Opinion in Chemical Engineering 2015, 9, 66-74.
- Floudas, C.A.; Niziolek, A. M.; Onel, O.; Matthews, L.R. Multi-Scale Systems Engineering for Energy and the Environment: Challenges and Opportunities. AIChE Journal 2016, 62 (3), 602-623.
- The Texas A&M Energy Price Index: A Predictive Framework, 2019
- Energy Information Administration (EIA), Electric Power Monthly, 2019
- U.S. Department of Energy (DOE), Green Power Markets, 2018

566273 Techno-Economic Analysis of Septic Tank Waste Treatment in a Centralized Hydrothermal Treatment Facility

Hydrothermal treatment, a thermochemical conversion process that is particularly useful for wet wastes, of septage produces valuable solid and liquid fertilizers while eliminating human pathogens. Additionally, it treats a problematic waste that is a burden for most traditional waste water treatment facilities. Operation of a centralized treatment facility would augment an area's capacity to utilize the nutrients and valuable products available in septage while also mitigating its potential environmental hazards. This facility would be alternative to landfilling, which can be detrimental to public waterways, or treatment at a traditional waste water plant. This analysis investigates the most important factors for an economically feasible facility that both treats septic tank waste and produces fertilizer products. The environmental impact of the plant is also compared to the standard methods of septage treatment.

578172 Turning a Hazard into Sustainable Revenue Streams: H_2S Valorization in Offshore Oil and Gas Productions through Process Intensification

The grand challenges of the 21st century lie squarely at the interplay of water, energy, and food— the so-called water-energy-food nexus. Add climate change into the mix, and the challenges become *more* challenging. Aggressive measures to decarbonize the energy market are necessary to tackle the threat of climate change. However, due to the ever-increasing global population, energy requirements are expected to increase 50% by 2030.[i] While ensuring that we protect the natural environment, it is also important to provide energy security. This calls for an energy supply that is consistent, reliable, sufficient, and affordable. Natural gas has arguably the least carbon footprint among fossil energy resources (carbon dioxide emission of burning natural gas is approximately half that of coal). While natural gas cannot be viewed as the permanent solution to energy security, it plays a vital role as the bridge fuel in the transition to a low-carbon economy.

Hydrogen sulfide (H_2S) is a major hazardous by-product in oil and gas production as well as many bio-based manufacturing processes, such as anaerobic digestion. Desulfurization of sour gas is essential prior to transportation or further processing due to its extremely toxic, acidic, and corrosive nature. Several offshore oil and gas production platforms are suffering from elevated levels of H_2S in recent years, and the production facilities are not designed for such high levels of H_2S. Currently, H_2S is removed using a non-regenerative scavenging process. The scavenger reacts with H_2S and produces less acidic, water-soluble compounds in an irreversible reaction.

Although the current process is rather established, it comes with significant challenges. Such challenges include high OPEX due to excessive amounts of the scavenger used in offshore productions, environmental concerns coming from the increasing levels of H_2S and increased mass of discharged spent scavenger, and fouling issues creating blockages in several downstream unit operations, such as wastewater treatment facilities.

For these reasons, design and development of more sustainable approaches that can fulfil the H_2S content threshold limit requirements and that can eliminate the disadvantages of the current non-regenerative H_2S-scavenging process are deemed necessary, specifically from an economic point of view. This work focuses on the design and techno-economic analysis of an intensified process that can be installed on an offshore platform to simultaneously remove and convert H_2S from produced natural gas to value-added product(s) to generate new revenue streams. The added benefits of this new process are that it allows for the efficient recovery of resources by turning the hazard, H_2S, into an opportunity. The proposed process is therefore well suited for the application of intensification concepts, integrating reaction and separation into a single unit.

[i] IEA, "World Energy Outlook 2012," International Energy Agency, 2012.

554760 Critical CO_2 Carnot Engine for Industrial Waste Heat Recovery and Utilization

Industrial processes regularly discard large volumes of low-grade waste heat to the environment. Harnessing this heat and transforming it into usable power would be environmentally and economically advantageous for industries. An overlooked solution to utilizing this waste heat is the use of a Carnot cycle to generate work with supercritical CO_2 as the working fluid. Supercritical $CO_2$is a novel working fluid for a Carnot engine as its critical temperature (31.1C) is much lower than the boiling point of water. This allows for the lower isotherm of the Carnot cycle to be constructed at the critical temperature, resulting in a sizable Carnot cycle with a near flat lower isotherm and upper isotherm just below the temperature of the low-grade heat. Thus, this type of Carnot cycle encompasses a larger area on a P-V diagram than any other thermodynamic conditions, thereby optimizing the amount of work that can be converted to power using a relatively small engine.We describe this approach of waste heat harnessing and illustrate ideas for utilizing the waste heat from an industrial toluene process. The impact of compression ratio, upper isotherm temperature, and pressure in the engine are investigated. Resulting savings amount to hundreds of thousands of kWh/year, thereby reducing cooler loads and electricity bills for the process, as well as reducing environmental emissions.

References:

[1] J.T. Banchero, B.D. Smith (Ed.), R.J. Hengstebeck. *Disproportionation of toluene*, 1969.

[2] Eldred H. Chimowitz Madeleine R. Laitz, F. Douglas Kelley. *Critical CO_2Carnot Cycle for Waste Heat Utilization*, 2017.

INDUSTRIAL SYMBIOSIS IN TRINIDAD AND TOBAGO: A FOCUS ON CARBON DIOXIDE

Thérèse G. Lee Chan and David A. Janes, The University of the West Indies, St. Augustine, Trinidad and Tobago

Abstract

Enterprise input-output (EIO) models are used to analyze the flows of resources, such as energy and materials, around industrial networks. The majority of the greenhouse gas, carbon dioxide (CO_2), emitted in Trinidad and Tobago (T&T) is from industrial sources. In this study, a simple representative model was developed of a network of CO_2 exchanges within an industrial park in T&T. The EIO model was developed from both process engineering and economic data. It incorporated performance indicators, which had been proposed in the literature, to calculate the level of industrial symbiosis (IS) and benefits - based on "the three pillars of sustainability" - realized in the representative industrial network. The eco-connectance, which is one measure of the level of IS, was determined as 1.3 for the representative network, thereby indicating the presence and level of IS in the industrial park. Furthermore, optimized flow networks were created, with and without a CO_2-reusing propylene carbonate plant. Multi-objective optimization was used to determine how to add the CO_2-reusing plant to the industrial network to minimize both CO_2 emissions and implementation costs. The addition of a CO_2-reuse plant reduced the CO_2 emissions by 3%, demonstrating there is scope for improving the existing IS network.

Introduction

Industrial symbiosis (IS) can be regarded as using, recovering and redirecting resource-laden wastes for reuse, in a network of separate commercial operations, so resources remain in productive use for longer. IS is considered an effective approach to waste management (Yazan, Romano et al. 2016) delivering both economic and environmental benefits. The twin-island state of Trinidad and Tobago (T&T) is rich in hydrocarbon resources, and was amongst those countries generating the highest carbon dioxide (CO_2) emissions per capita at 25.72 tonnes per head in 2016 (Knoema World Data Atlas, 2018). In 2014, 74% of the CO_2 emitted in T&T originated from industry (UNDP, 2014). The majority of the petrochemical industry in Trinidad is housed in the Point Lisas Industrial Estate (PLIE), which contains several ammonia and methanol plants.

In this paper, the exchanges of process CO_2 between various petrochemical plants in PLIE are used to assess the level of industrial symbiosis occurring on the industrial park. The aim of this work was to develop a simple representative model of this nascent symbiotic industrial system, in terms of CO_2 exchanges at PLIE. Our model was then used to perform an integrated process engineering and economic analysis to quantify the level of exchanges and benefits beneath "the three pillars of sustainability" and thence determine the scope for improvement.

In part because there are currently no legislative restrictions on CO_2 emissions in T&T, PLIE is not thought of locally as being a bastion of progressive industrial ecology. Although PLIE was not originally planned for industrial symbiosis, waste CO_2-streams have been shared amongst plants since 1983 (Boyle and Baetz, 1998) when the Trinidad and Tobago Urea Company plant was commissioned to use waste process CO_2, ammonia and some facilities from a nearby separately-run Fertrin ammonia plant to produce urea. The current "self-organized" state of sharing of waste streams developed over the years, rather like early sharing at archetypical eco-industrial park (EIP) at Kalundborg, Denmark (Ehrenfeld and Gertler, 1997), in that each additional exchange in the park was an independently negotiated business deal, which was established only if expected to be commercially beneficial.

Few studies have been conducted on the application of industrial symbiosis in T&T. In 1998, four companies in PLIE were evaluated in terms of waste management and the study (Boyle and Baetz 1998) found that there was potential for the waste reduction by reusing and recycling, although the study did not consider CO_2 as a waste resource. In 2008, Boodlal (2008) compiled an inventory of CO_2 resources in Trinidad, in which the emissions from the chemical plants at PLIE were calculated with appropriate stoichiometric ratios for the relevant reactions.

Methodology

There is no standard methodology for measuring the level of IS in an industrial network or park. Some authors use economic, environmental and social indicators to show how IS improves a park by comparing the indicators for the park with and without IS activities. However, this does not give an actual measure of the level of IS. Some of the proposed measures for industrial symbiosis are eco-connectance, by-product and recycling rate (Tiejun 2010), eco-efficiency (Park and Behera 2014), environmental impact momentum (Yilmaz et al., 2016) and technical exchange efficiency (Fraccascia, Albino et al., 2017). Yilmaz et al. (2016) published a comprehensive list of indicators that can be used when considering industrial symbiosis.

Material flow analysis (MFA), enterprise input-output (EIO) tables, life cycle analysis (LCA), eco-efficiency, eco-connectance and waste recovery are some of the tools which have been used in reducing waste in a system in a sustainable manner. MFA, LCA and eco-efficiency are not preferred since they do not consider the whole picture.

In this study, the process CO_2 emissions for selected plants were evaluated based on reported information about the various technologies employed and the plant capacity. Flue-gas CO_2 emissions were excluded from this study, because of the much lower CO_2-purity of these streams. Potential means for improving the sharing and reusing of CO_2 to create shared value, that is the IS level, were considered. Nine petrochemical plants in PLIE were selected as being representative of the industrial park. The plants considered were ammonia, methanol, urea and melamine. An EIO table for the industrial park was developed in MATLAB® (ver. R2019a), the chosen modeling environment.

The general equations and variables for an EIO system are:

$$X = Ax + f \tag{1}$$

$$r = Rx \tag{2}$$

$$w = Wx \tag{3}$$

Variables:

A – Technical coefficient matrix
x – Total output vector
f – Vector of final demands
R – Resource intensity matrix
r – Vector of primary input
W – Waste intensity matrix
w – By-products and waste vector

It is assumed that there is only one main product for each process. A, R and W are matrices of constants and are typically observed or calculated values for a specified technology for the process. Therefore, if process demand is changed, the outputs from the process would be changed by that value.

Three EIO tables were developed for a *base*, *current* and *proposed* case. The base case was one in which there was no symbiosis occurring and this was used as a baseline in the study. In the current case, the existing symbiosis was shown and in the proposed case, a CO_2-reusing plant was added to the industrial park. Material and energy flows for the base and current cases were estimated on the basis of the average production rate of the various plants over a three-year period. Various plants, which reuse CO_2, were considered, such as manufacturing processes for biofuels (Benn-Ogeer et al., 2015), propylene carbonate (Demirel, 2015) and polyethercarbonate polyols (von der Assen and Bardow, 2014). The propylene carbonate plant was selected based on the feasibility and the availability of published performance information. The data used to simulate and cost the plant in Aspen Plus® (ver. 10) was obtained from Demirel (2015). For all the cases, the direct and total coefficients were determined. Besides the evaluation of the direct and total coefficients, the level of industrial symbiosis occurring in the various scenarios was evaluated using various industrial symbiosis measures and sustainability indicators.

In this study, the focus was on the exchange of process CO_2 amongst the plants. The CO_2-rich process streams, which originated from the ammonia plants, were supplied to the sinks at an assumed CO_2-purity of 99 wt%. US dollars were used for all prices and costs. The price of CO_2 supplied to the sinks was determined from the calculation of the levelized cost of CO_2 transportation and an assumed raw material cost for CO_2, to buttress the mutualism of the IS. Consequently, pipelines from the CO_2 sources to sinks were evaluated. The equation developed by McCoy and Rubin (2008) was used to estimate the diameter of the pipelines required for the sink flows. The costs associated with purification and compression of the CO_2 were taken into consideration and together with the diameter and length of the pipeline, these were used in the determination of the levelized cost of CO_2 transportation (\$ $tonne^{-1}$ of CO_2).

Optimization of the industrial park was computed in MATLAB by taking into consideration the amount of waste CO_2 and the economics. Single objective optimization for each case was carried out to minimize waste CO_2 and to minimize the cost. Cases were considered using linear programming (LP) and mixed integer linear programming (MILP). In the MILP, binary variables were used to prevent small flow rates through the pipelines. For the LP and MILP cases, the dual-simplex algorithm and the branch and bound algorithm were respectively utilized. Finally, multi-objective optimization using the genetic algorithm in the MATLAB function *gamultiobj* was used to minimize the amount of waste CO_2 and the cost. The Pareto front for each case was generated.

Results and Discussion

Material and energy flows were calculated for the input and output of the plants in the industrial park. These flows were then used to generate the EIO tables for the base, current and proposed cases. The cases showed that from the ammonia plants, there were approximately 202,000 kg/h of nearly pure CO_2 being produced. The EIO for the current case indicated that approximately 35% of the CO_2 was reused in methanol and urea plants in the industrial park and the remainder was vented to the atmosphere. Introducing a new plant, which reused CO_2 from the industrial park, increased the CO_2 usage by approximately 3%, thereby reducing the amount of CO_2 emitted to the atmosphere. From the EIO table, the direct and total coefficients were determined. Analysis of the direct coefficients showed that the largest direct coefficient for the by-products and waste was attributed to the nearly pure CO_2 being emitted from the ammonia plants.

The core EIO model gives an idea of the material and flow exchanges within the network, but does not measure the level of industrial symbiosis. The calculated values for the performance indicators, used to measure IS in these three cases, are shown in Table 1.

Table 1. Industrial Symbiosis Performance Indicators

Industrial Symbiosis Indicator	Base	Current	Proposed
Eco-connectance	0	1.3	1.8
Environmental impact momentum	0	0.56	0.62
Fraction of waste CO_2 utilized	0	0.36	0.38

In the base case, it was assumed that there were no exchanges amongst the plants. The values obtained for the current case show that there is some level of industrial symbiosis occurring and those for the proposed case indicate that the addition of a plant which utilizes CO_2 as a raw material can lead to increased levels of industrial symbiosis.

In this study, a propylene carbonate plant was proposed for the industrial park. There are several ways in which CO_2 can be utilized, but because many of these uses are emerging technologies, it is difficult to obtain information for them. The propylene carbonate plant had a capacity of approximately 10,500 kg/h and consumed 4,500 kg/h of CO_2. This capacity was determined in accordance with the current market for propylene carbonate. The results from the model indicated that the feasibility of such a plant is highly dependent on the price of the catalyst since this was the most expensive component of the model.

One of the objectives of this study was to optimize the flow of CO_2 in the three cases. This optimization considered minimizing the waste CO_2 and the cost. The levelized cost of transporting the CO_2 was calculated and hence pipelines from the CO_2 sources to the sinks were considered. The transportation costs of CO_2 ranged from US$9 per tonne to $50 per tonne of CO_2. The higher end of the costs was attributed to smaller flows in the pipelines and hence due to a fixed cost recovery period, the price would consequently be higher. It should be noted that the costs obtained were derived from a fixed flow in the pipeline and did not take into account the change in the flow rate and the cost. This was done to ensure that the optimization model remained linear. According to literature, there is no fixed cost for CO_2 since it is dependent on the purity of CO_2, distance and the state in which the CO_2 is transported.

Various optimization cases were considered using the current and proposed data. In the first scenario, minimization of waste CO_2 was carried out. MILP and LP were used to solve the optimization models for the base and proposed cases. Results from the LP case with the current data indicated that the CO_2 from four out of the five plants should be sent to the CO_2 sinks. The MILP results indicated the same but the flows were different for the plants. However, for the proposed cases both MILP and LP showed that flows should be sent to the sinks from all the plants. It is suggested that some flows be sent from each plant since the existence of a plant venting all its CO_2 to the atmosphere would increase the environmental burden for that particular plant.

The optimal annual costs for the current and proposed cases were $12 million per year and $14 million per year respectively. It is evident that in the current case, the transportation cost for CO_2 is less since there are less sinks which require CO_2. It should also be noted that for the proposed case, there would be some additional expenditure in constructing and maintaining the propylene carbonate plant. Multi-objective optimization for minimizing the CO_2 waste and cost was also done for both cases. The annual costs for each case were $12.2 million per year and $14.6 million per year, respectively.

Sustainability indicators were employed to evaluate the optimized cases using multi-objective optimization (see Table 2). The eco-efficiency was determined as a ratio of the annual production to the amount of CO_2 used.

Table 2. Sustainability Indicators

Indicator	Sustainability Dimension	Units	Current Case	Proposed Case
Eco-efficiency	Environmental	-	9.8	9.4
CO_2 emissions	Environmental	t/h	277	273
Resource efficiency	Environmental	t/h	0	4.5
Relative change of CO_2 recycling	Environmental	%	36	38
Total Capital Investment (CO_2 transport)	Economic	$	22.5	26.5
Total Capital Investment (new plants)	Economic	Million $	-	17.2
Profit as a result of IS activities	Economic	Million $/y	0.41	(0.43+3.6)
Job Creation	Social	-	No	Yes
Creation of Industrial Symbiosis	Social	-	6	7

The new plant led to a reduction in the overall CO_2 emissions thereby increasing the amount of CO_2 being reused. Addition of the propylene carbonate plant resulted in a decrease in eco-efficiency. Nevertheless, it would lead to an increase in jobs and would create an increased level of industrial symbiosis in the industrial estate.

Further work may investigate the actual linkages amongst the plants and determine the capacities of the existing pipelines. It could also consider looking at CO_2 in flue gas and other CO_2 utilization options. Additionally, exchanges of other materials and energy from the industrial park need to be included to get a fuller picture of how industrial symbiosis is being and could be put into practice.

Conclusions

The model indicates that current partners in the on-going industrial symbiosis at the Point Lisas Industrial Estate ought to be achieving sustainable benefits. The estimated levelized cost for CO_2 transportation ranged from $9 per tonne to $50 per tonne of CO_2. The introduction of a few chemical manufacturing plants reusing CO_2 should lead to increased IS indicator values and further triple-bottom-line benefits, even though it would only make a small dent in the quantity of as yet unused process carbon dioxide. The integration of data from process engineering models into EIO models provides a means to analyze and optimize potential IS scenarios.

References

1. Benn-Ogeer S. K., D. A. Janes and R. Deo (2015). "A Life Cycle Assessment (LCA) of the Production of Biofuels in Trinidad from Waste CO_2 With Microalgae", e-Journal of the Caribbean Academy of Sciences 8(1).

2. Boodlal, D. (2008). "Trinidad and Tobago's CO_2 Inventory and Techno-economic Evaluation of Carbon Capture Options for Emission Mitigation." 3rd Tobago Gas Technology Conference, Lowlands, Tobago.

3. Boyle, C. A. and B. W. Baetz (1998). "A prototype knowledge-based decision support system for industrial waste management: Part II. Application to a Trinidadian industrial estate case study." Waste Management 17(7): 411-428.

4. Demirel, Y. (2015). "Sustainability and Economic Analysis of Propylene Carbonate and Polypropylene Carbonate Production Processes Using CO_2 and Propylene Oxide." Chemical Engineering & Process Technology 0: 6:3.

5. Ehrenfeld, John; Nicholas Gertler (1997). "Industrial Ecology in Practice: The Evolution of Interdependence at Kalundborg". Journal of Industrial Ecology 1: 67–79.

6. Fraccascia, L., V. Albino and C. A. Garavelli (2017). "Technical efficiency measures of industrial symbiosis networks using enterprise input-output analysis." International Journal of Production Economics 183: 273-286.

7. Knoema World Data Atlas. (2018). "Trinidad and Tobago – CO_2 emissions per capita". [retrieved 01/04/2019]. Available from: https://knoema.com/atlas/Trinidad-and-Tobago/CO2-emissions-per-capita.

8. McCoy, S. T. and E. S. Rubin (2008). "An engineering-economic model of pipeline transport of CO_2 with application to carbon capture and storage." International Journal of Greenhouse Gas Control 2(2): 219-229.

9. Park, H. S. and S. K. Behera (2014). "Methodological aspects of applying eco-efficiency indicators to industrial symbiosis networks." Journal of Cleaner Production 64: 478-485.

10. Tiejun, D. (2010). "Two quantitative indices for the planning and evaluation of eco-industrial parks." Resources, Conservation and Recycling 54(7): 442-448.

11. UNDP, Factor CO_2. (2014). "Projection of the Business as Usual (BaU) and measures scenarios of carbon emissions for the electrical energy generation, industry and transportation sectors". Port of Spain, Trinidad and Tobago.

12. von der Assen, N., and A., Bardow, (2014). "Life cycle assessment of polyols for polyurethane production using CO_2 as feedstock: insights from an industrial case study." Green Chemistry 16: 3272-3280.

13. Yazan, D. M., V. A. Romano and V. Albino (2016). "The design of industrial symbiosis: an input–output approach." Journal of Cleaner Production 129: 537-547.

14. Yilmaz, O., E. Yontem and E. Alkaya (2016). Industrial Symbiosis Indicators. FISSAC.

559393 Thermal Stability of Imidazolium-Based Ionic Liquids: The Effects of Alkyl Chain Length, Environment, and Heating Conditions

Ionic liquids are room-temperature molten salts that have found increasing use as solvents for polymer and biomaterial processing. This study investigates a series of 1-alkyl-3-methylimidazolium ionic liquids, which have been found to effectively dissolve lignocellulosic biomass. This family of ionic liquids has received significant attention in the literature. Despite this, the thermal stability and degradation behavior of these solvents is not well understood. Furthermore, there have been only a few investigations into the effect of structural variations such as varying the length of their alkyl chains.

In the present study the thermal stabilities of these ionic liquids were investigated with the ethyl, propyl, butyl, hexyl, and decyl alkyl chains, and the chloride and acetate cations. The ionic liquids were studied under both isothermal conditions from 100 to 200 ^{o}C over sixteen hours, and under dynamic heating conditions with heating rates from 5 to 20 ^{o}C/min up to 650 ^{o}C. It was found that the ionic liquids with longer alkyl chains generally displayed decreased thermal stability.

The effect of the environment on the stability of ionic liquids is also significant, and was investigated with 1-ethyl- and 1-butyl-3-methylimidazolium acetate in mixtures with ethanol, methanol, water, and *N,N*-Dimethylformamide (DMF) at elevated temperatures and pressures, up to 150 ^{o}C and 350 bar. A custom high-pressure view cell was used in conjunction with a UV-vis spectrophotometer to capture the degradation of the ionic liquid-solvent mixtures over time. It was found that in protic solvents, including ethanol, methanol, and water degradation was suppressed as reflected by the UV absorption patterns. In contrast DMF, an aprotic solvent, was found to significantly accelerate thermal degradation.

This research is supported by NSF Grant CBET-1509390.

558616 Thermodynamics of the Mixtures of Ammonia and Ionic Liquids

Ionic liquids (ILs) are liquid salts composed entirely of ions, and defined to have a melting point below 373 K. Ammonia (NH_3) is one of the most important and largest volume industrial chemicals produced today. The fundamental thermodynamic knowledge for ILs with NH_3 is essential for exploring the feasibility of ILs in various applications such as gas separation and absorption-refrigeration. This study is divided into three parts. In the first part, the vapor-liquid-equilibria of NH_3 in ILs will be reported. The solubility of NH_3 in ILs was measured using the gravimetric method (XEMIS Microbalance, Hiden Iscohema, UK) at pressures up to 0.7 MPa and at temperatures of 283.15, 298.15, 323.15 and 348.15 K. Experimental vapor-liquid-equilibria data was modeled using Peng-Robinson Equation of State and Non-Random Two Liquid (NRTL) activity coefficient models. In the second part, the direct experimental determination of enthalpies of solution of NH_3 in ILs using high pressure Calvet calorimeter (SETERAM BT 2.15, France) will be reported. Heat of solution is a fundamental thermodynamic property that is related to the nature of interaction between the gas and the IL. The enthalpy of NH_3 absorption in IL can reflect the heat effect due to physical dissolution of NH_3 in the IL and heat effect due to chemical interaction between the two if present. In the third and final part, proton Nuclear Magnetic Resonance (1H NMR) technique will be reported to elucidate the interactions between NH_3 and ILs.

563947 Characterization of 1-Ethyl-3-Methyl-Imidazolium Ionic Liquid Behavior in Non-Aqueous Solvents

We characterize binary systems of 1-Ethyl-3-methyl-imidazolium $[emim]^+$ ionic liquids (ILs) with various molecular solvents including; glycerol, ethylene glycol, acetonitrile, and triethanolamine using precise measurement techniques to determine the conductivity, viscosity, and density of these systems across a wide IL concentration range. We use these three properties in conjunction with the estimated Stoke's radii of the ions calculated from the ubiquitous UNIQUAC and UNIFAC and values, to estimate the ion dissociation based on a method previously described in our previous work.[1,2] Additionally we explore the role of hydrogen bonding in these systems by measuring excess enthalpy across a wide range of IL concentrations. These measurements offer new insights into the driving forces behind ion pair formation in IL/solvent systems. We further consider the solvent effects on ion pair formation, by studying solvents with similar dielectric constants and vary structure (i.e. glycerol, triethanolamine, and ethylene glycol) as well as solvents with similar structure and different dielectric constants (i.e. triethanolamine and glycerol).

(1) Nordness, O.; Simoni, L. D.; Stadtherr, M. A.; Brennecke, J. F. Characterization of Aqueous 1-Ethyl-3-Methylimidazolium Ionic Liquids for Calculation of Ion Dissociation. *J. Phys. Chem. B* **2019**, *123*, 1348–1358. https://doi.org/10.1021/acs.jpcb.8b11892.

(2) Simoni, L. D. Predictive Modeling of Fluid Phase Equilibria for Systems Containing Ionic Liquids, University of Notre Dame, Indiana, 2009.

568983 Investigation of Ion Correlations in Ionic Liquid and Organic Solvent Mixtures through Scalable Screening

Investigation of Ion Correlations in Ionic Liquid and Organic Solvent Mixtures Through Scalable Screening

Ray A. Matsumoto[1,2], Matthew W. Thompson[1,2], and Peter T. Cummings[1,2]

[1]*Multiscale Modeling and Simulation Center, Vanderbilt University, Nashville, TN, 37235-1604, USA*

[2]*Department of Chemical and Biomolecular Engineering, Vanderbilt University, Nashville, TN, 37235-1604, USA*

ABSTRACT

Considerable efforts have been made to improve the performance of supercapacitors, electrochemical devices that exhibit high power density but low energy density. As part of these efforts, room temperature ionic liquids have shown to improve the energy density of supercapacitors due to their wide operational voltage windows in comparison to other electrolytes[1]. Ionic liquids however exhibit high viscosities and slow dynamic properties which can negatively affect the charging rates of such devices. To mitigate slow dynamics, ionic liquids can be solvated in organic solvents without a substantial loss in operational voltage window[2].

We have previously used a screening approach to encompass a larger chemical space and expand our knowledge of interactions between ionic liquids and organic solvents. As a result, we implemented the necessary algorithms and force fields within MoSDeF[3] to conduct screening of $[BMIM^+][Tf2N^-]$ in 22 distinct solvents over a range of 18 different concentrations. Self-diffusivities of ions in these solutions were computed via mean squared displacement (MSD) calculations, while conductivity was estimated via the Nernst-Einstein (NE) equation. From the ability to sample a larger portion of the solvent-ionic liquid design space, we were able to identify several key trends related to ionic liquid diffusivity and conductivity.

Using the previous screening parameter space, we now focus our efforts towards understanding the structure of these ionic liquid and organic solvent mixtures. Ion correlations have been well-characterized for bulk or so-called "neat" ionic liquids, which have shown to impact conductivity through the concept of "ionicity" [4]. On the other hand, ion correlations in ionic liquid and solvent mixtures are still relatively unknown and therefore, we characterize ion correlations within these ionic liquid and organic solvent mixtures by analyzing ion pairing and caging. Furthermore, we compare the Einstein-Helfand (EH) approach[5] to the NE equation for estimating the ionic conductivity of these solutions. Unlike the NE equation, the EH formalism considers ion correlations, which enables us to investigate ionicity through a ratio of the two conductivities. A greater understanding of the relationship between ion correlations and dynamics presented can further guide the selection of solvents and concentration for electrochemical applications of ionic liquids.

References

[1] L. P. Rebelo, J. N. Lopes, J. M. Esperanca, E. Filipe, "On the Critical Temperature, Normal Boiling Point, and Vapor Pressure of Ionic Liquids", Journal of Physical Chemistry B, **2005**, 109, 6040-6043

[2] R. Lin, P. Huang, J. Segalini, C. Largeot, P.L. Taberna, J. Chmiola, Y. Gogotsi, P. Simon, "Solvent Effect on the Ion Adsorption from Ionic Liquid Electrolyte Into Sub-nanometer Carbon Pores", Electrochimica Acta,**2009**, 54, 7025-7032.

[3] "MoSDeF" [Online]. Available: https://github.com/mosdef-hub.

[4] D. R. MacFarlane. *et al.*, "On the Concept of Ionicity in Ionic Liquids", Physical Chemistry Chemical Physics, **2009**, 11, 4962-4967

[5] C. Schroder, M. Haberler, O. Steinhauser, "On the Computation and Contribution of Conductivity in Molecular Ionic Liquids", The Journal of Chemical Physics, **2008**, 128, 134501

573482 *In Situ* Grazing-Incidence Wide Angle X-Ray Scattering (GIWAXS) Investigation of Effects of Silica Nanopore Confinement on Crystallization of [BMIM] Ionic Liquids

Supported ionic liquids (ILs) in nanoporous matrices are being developed to overcome the cost and unknown toxicity of bulk ILs, and to accentuate the microenvironments they offer for separation and catalysis. However, it is not yet known to what extent confinement into nanometer-scale pores alters the surface-induced ordering of ILs. The present work gives insights into how silica mesopore confinement affects the crystallization behavior of two selected ILs based on *in situ* GIWAXS experiments performed at Argonne National Laboratory's Advanced Photon Source.

1-Butyl-3-methylimidazolium hexafluorophosphate ([BMIM]PF_6) and [BMIM]Cl are among the most extensively studied ionic liquids (ILs), which makes bulk crystallization information available for comparison. Silica films with orthogonally oriented 10 nm mesopores are prepared by templating with Pluronic surfactant P123. Some films are modified by covalently tethering 1-(3-trimethoxysilylpropyl)3-methylimidazolium chloride [TMS-MIM]Cl onto the pore wall and XPS is used to verify uniform modification throughout the depth of the films. Both ILs underwent cold crystallization below their glass transition during heating from low temperatures. GIWAXS patterns were collected *in situ* during heating. The crystal phase transition temperatures and corresponding crystal phases were captured and compared for bulk ILs and confined ILs in silica nanoconfinement with and without tethering. Confinement of both ILs in unmodified silica mesopores causes changes in their crystal phases and phase transition temperatures compared with bulk ILs. The melting point, where the long range order completely disappears from the GIWAXS pattern, of [BMIM]PF_6 is about 17.5 °C higher with confinement compared with bulk [BMIM]PF_6 while it is about 190 °C lower for [BMIM]Cl. Confined [BMIM]PF_6 in silica mesopores tethered with [TMS-MIM][Cl] shows a stable and symmetric crystal structure which did not melt at room temperature even though the bulk IL as a film on a Si wafer melts at -11 °C. The melting point depression for [BMIM]Cl is consistent with the Gibbs-Thomson equation with a surface energy of 8.0 mN/m. However, the melting point increases for [BMIM]PF_6 cannot explained by capillary effects. The interaction between ILs and the silica surface (most likely [BMIM] interaction with the negatively charged surface) and molecular rearrangement due to nanoconfinement are believed to be the main reasons for the changes observed. These findings show that when ILs are confined, drastic changes in structure can occur that alter the liquid-state temperature operating range for applications such as ion conduction and supported catalysts.

566210 Impact of Carbon Number of Carboxylic Acid on the Polarity of Hydrophobic Deep Eutectic Solvent

Solvatochromic dyes offer a powerful look into the molecular forces that exist in solutions. Zwitterionic dyes are particularly useful for the investigation of hydrophobic deep eutectic solvents (HDES), which interact via a hydrogen bond. This bond results in a solution with a melting point significantly lower than that of either hydrogen bond acceptor (typically a quaternary ammonium salt) or donator. The hydrogen bond donor, which is usually a carboxylic acid for hydrophobic deep eutectic solvents, is extremely influential on this melting point, with octanoic acid based HDES being liquid at room temperature and dodecanoic acid HDES being a solid until 35-45 °C. By taking advantage of this phenomena, the hydrogen bond acceptor can be changed to impact physical properties like Henry's law constants without dramatically changing polarity or melting point of the solvent. COSMOTherm, a density functional theory based molecular modeling software, is used to explore the impact of carboxylic acid length on the sensitivity of HDES systems to change in N-quaternary ammonium bromide salt characteristics. These results are verified using solvatochromic dyes 4-nitroaniline and 4,4 N-diethyl 4-nitroaniline.

575714 The Effect of Cation Polarity on the Melting Points of Ionic Liquids: An Experimental and Computational Study

A number of molten salts cannot be classified as ionic liquids (ILs) because they are not liquids at or below 100 C. These materials may have interesting or desirable properties however because they remain solids at standard temperatures this is likely to severely restrict their usage in a given process. Lowering the melting points of these materials without significantly affecting their other properties, effectively extending their liquidus range, could facilitate their adoption in industrial processes.

Recently, it was shown that the addition of different substituents to a series of thermally-robust IL cations could have a very substantial effect on the melting point. It was concluded that electron withdrawing effects in these groups create polarity within the cations, which could align, lowering the repulsion in these systems and hence the liquid phase enthalpy, leading to a depression in the melting point. From these results, it remained unclear if this principle was system specific or if it could be reasonably extended to other IL classes.

In this talk we present a systematic study of single fluorine substitutions on the structure and thermodynamics of different homologous series of IL cations. We systematically vary the position of the fluorine substitution in perarylsulfonium, perarylphosphonium, imidazolium, and pyridinium cations paired with bistriflimide, effectively altering the electronic charge and dipole moment of these cations. These ILs were synthesized and their melting points measured along with their enthalpies and entropies of fusion. Furthermore, the crystal structures of the compounds were determined using X-ray crystallography and used to simulate these structures using molecular dynamics. Using two-state simulations, we are able to predict the melting points of these ionic liquids and connect the effect that induced polarity in the IL cation has on the overall melting points.

565090 Change Regulation of Cellulose Depolymerization during Dissolving Process in Ionic Liquids

Abstract

Regenerated cellulose fibers are commercially produced by viscose process using cellulose as raw materials. However, this process exits some drawbacks including high cost, solvent volatility and toxicity. Therefore, it is necessary to develop more efficient and greener technology for cellulose dissolution. Ionic liquids (ILs), as green solvents with attractive properties of non-volatility, thermal stability, and tunable structure, show excellent solubility in dissolving cellulose and spinning, which has been regarded as a promising method for cellulose solution.

However, cellulose degradation is inevitable in the process of cellulose dissolution in ILs, which significantly influences the mechanical properties of regenerated fibers. The mechanical properties of regenerated fiber decreased approximately linearly with cellulose depolymerized. Thus, it's vital to understand the change regulation of cellulose depolymerization in ILs, and hence guides to obtain regenerated fibers with high tensile strength. Usually, the reduction of degree of polymerization (DP) is used to evaluate the degree of cellulose depolymerization.

In this work, microcrystalline cellulose and wood pulp were dissolved respectively in two ILs, i.e., 1-allyl-3-methylimidazolium chloride (AmimCl) and 1-ethyl-3-methylimidazolium diethyl phosphate (EmimDep). The cellulose depolymerization was investigated within 72 hours at 90~130°C. The results indicate that the DP of regenerated cellulose decrease with time, and the decrease in DP of regenerated cellulose became more pronounced when the temperature is above 100°C. And cellulose depolymerization in EmimDep rather than AmimCl has a slight change with the increasing of reaction temperature and time. Therefore, high-quality regenerated cellulose fiber will be acquired when using EmimDep as solvent and no more than 100°C.

Reference

1. Liu, Y. R., K. Thomsen, Y. Nie, S. J. Zhang & A. S. Meyer (2016) Predictive screening of ionic liquids for dissolving cellulose and experimental verification. *Green Chemistry,* 18, 6246-6254.
2. Douglass, E. F., H. Avci, R. Boy, O. J. Rojas & R. Kotek (2018) A Review of Cellulose and Cellulose Blends for Preparation of Bio-derived and Conventional Membranes, Nanostructured Thin Films, and Composites. *Polymer Reviews,* 58, 102-163.
3. Barthel, S. & T. Heinze (2006) Acylation and carbanilation of cellulose in ionic liquids. *Green Chemistry,* 8, 301-306.

571838 Discovery of New Ionic Liquids Via Molecular Simulations for the Separation of Azeotropic Mixtures of High Global Warming Potential Hydrofluorocarbon Refrigerants

In this work, we use molecular simulations to discover new ionic liquid (IL) separating agents for azeotropic separation of hydrofluorocarbon (HFC) mixtures. Mandated by the 1987 Montreal Protocol, chlorofluorocarbon (CFC) refrigerants have been gradually replaced by HFCs to prevent ozone depletion. Many of these second-generation HFC mixtures, however, have a high global warming potential (GWP) and the 2016 Kigali agreement ordered their gradual phase-out [1]. Due to the often azeotropic compositions of these HFC refrigerants, existing separation methods for removing the high GWP components are currently infeasible or not practical, but it is wasteful to incinerate HFC mixtures as some HFC components have low GWPs and can be recycled. We hypothesize that custom ILs can be designed to remove low GWP HFC components from specific HFC mixtures [2][3]. However, because millions of potential ILs exist [4], prediction of IL properties via molecular simulations is necessary to screen for process-specific ILs prior to their synthesis. To correctly simulate IL-HFC systems, molecular simulations require accurate force field parameters for the molecules being simulated. To this end, molecular simulations were performed to determine thermophysical properties, including phase equilibria, of HFCs using the Cassandra Monte Carlo modeling package. Separate simulations were run using parameters from three force fields, Generalized Amber Force Field (GAFF), Optimized Potentials for Liquid Simulations (OPLS), and a customized force field. The critical temperatures and pressures of the HFCs were calculated using the traditional method of rectilinear diameters. The resulting critical properties for each force field were compared to experimental thermodynamic data to determine optimal force field parameters for the HFC molecules. We then quantified the uncertainty in the critical property predictions to study how the uncertainty propagates to single-component phase equilibria calculations [5]. This uncertainty analysis and comparison with experimental data offers quantitative insights to the suitability of molecular simulations to screen ILs for further study as HFC separating agents.

References

[1]United Nations Environment Programme. Ozone Secretariat. (2006). Handbook for the Montreal protocol on substances that deplete the ozone layer. UNEP/Earthprint.

[2]Plechkova, N. V., & Seddon, K. R. (2008). Applications of ionic liquids in the chemical industry. *Chemical Society Reviews*, *37*(1), 123-150.

[3]Chávez-Islas, L. M., Vasquez-Medrano, R., & Flores-Tlacuahuac, A. (2011). Optimal molecular design of ionic liquids for high-purity bioethanol production. *Industrial & Engineering Chemistry Research*, *50*(9), 5153-5168.

[4]Holbrey, J. D., & Seddon, K. R. (1999). Ionic liquids. *Clean products and processes*, *1*(4), 223-236.

[5] Hajipour, S., & Satyro, M. A. (2011). Uncertainty analysis applied to thermodynamic models and process design–1. Pure components. *Fluid Phase Equilibria*, *307*(1), 78-94.

578136 Commercial Applications of Ionic Liquids

Several processes and products have been developed using ionic liquids (ILs) in the past decade; however, the general belief is ILs are only academic curiosities and are too expensive to use in commercial applications. This presentation which is based on a new book entitled "Commercial Applications of Ionic Liquids" will highlight several applications using ionic liquids. A few examples which will be highlighted in this presentation include: (1) a new class of capillary gas chromatography (GC) columns with stationary phases based on ILs which have been developed by Professor Daniel Armstrong at the University of Texas in Arlington; (2) the development of a new chloroaluminate IL-based alkylation catalyst by Chevron as an alternative to using corrosive and toxic hydrofluoric acid (HF) as a catalyst; (3) a new process for electrospinning bionrenewable chitin sorbents for uranium extraction from seawater using ILs developed by Professor Robin Rogers and 525 Solutions; (4) the deconstruction of biomass using ILs prepared by Jason Hallett at Imperial College; and (5) the development of supported ionic liquid phase (SILP) materials for catalysis by Prof. Dr. Peter Wasserscheid at the Universitat Erlangen, Germany. In all over 50 applications have been developed at the pilot or commercial scale and companies manufacturing ionic liquids will also be highlighted.

564783 Whole-Cell Bioconversion of Phytosterols in Ionic Liquids Co-Solvents with Alleviation of Product Inhibition

Androst-4-ene-3,17-dione (AD) obtained by sterols side-chain cleavage is a vital intermediate of steroids. However, product (AD) inhibition and low substrate (sterols) water-solubility limit the further improvement of the bioconversion efficiency. For these two reasons, ionic liquids (ILs) capable of dissolving many insoluble substrates and poly (2-hydroxyethyl methacrylate) (pHEMA) cryogel microbeads with ability to adsorb AD were used in this work for sustainable bioconversion of phytosterols to produce AD.

First, cholinium amino acids ([Ch][AA]) ILs with good biocompatibility were introduced as co-solvents in phytosterols bioconversion by *Mycobacterium* sp. MB 3683 resting cells. The physicochemical properties (density, viscosity, hydrophobicity and polarity) of [Ch][AA] ILs were investigated based on their chemical structures. The solubility of phytosterols and AD in [Ch][AA] ILs aqueous solution was measured. The results indicated that [Ch][AA] ILs especially [Ch][Glu], [Ch][Asp] and [Ch][Gln] showed proper biocompatibility to *Mycobacterium* sp. MB 3683 cells and the production of AD in 1% [Ch][Asp] was about twice as that in the buffer. Moreover, pHEMA cryogel microbeads with pore size of 1~50 μm were added to the fermentation process of phytosterols by *Mycobacterium* sp. MB 3683. Obviously, pHEMA cryogel microbeads were found to promote the bioconversion to produce AD due to their high capacity of adsorption to AD. The effects of dry cell weights, phytosterols concentration on AD production, and the synergy interaction of pHEMA cryogel microbeads and preferred ILs were investigated in detail.

In situ removal of AD by pHEMA cryogel microbeads reduced the toxicity of products to cells and thus the production of AD was increased, as well as the subsequent separation process was simplified. Thus, by the addition of [Ch][AA] ILs and pHEMA cryogel microbeads respectively, two problems that limit AD production have been effectively solved, which is potential for microbial side-chain cleavage of phytosterols and up-scale AD production.

Keywords: Cholinium amino acids ionic liquids; pHEMA cryogel microbeads; Bioconversion; Phytosterols

576714 New Application of Ionic Liquids As a Sweep FLUID in Methanol Synthesis in a Membrane Reactor

Methanol (MeOH) finds broad use as a transportation fuel and as an important raw material used in the production of other fuels and chemicals. There are several commercial methanol synthesis (MeS) processes (e.g., Lurgi and Haldor-Topsoe), but they are all challenged by low per-pass conversion that necessitates the recycling of the unreacted syngas. This challenge is especially problematic for MeOH production from renewable biomass, where the use of oxygen-blown gasifiers is not economic, and the available syngas has a large nitrogen content. One way to overcome the thermodynamic limitations of the MeS reaction, and thus increase the per-pass conversion is to utilize a membrane reactor (MR). In this study, a novel high-pressure MR is used, in which a membrane with the desired characteristics serves as an interface contactor between the MeS environment in the shell-side and a sweep liquid solvent flow in the membrane permeate-side. Regarding the sweep liquid, the application of an ionic liquid (IL) is investigated in the MR in order to separate methanol during MeS with higher efficiency. The general idea is to increase the rate of conversion of the syngas into MeOH via its in-situ separation through a circulating IL solvent. Specifically, the MeOH synthesized from dilute syngas, typical of that produced from air-blown biomass gasifiers, is selectively removed in-situ via membrane separation, using an IL solvent purge (sweep).

First, in this work, the solubilities of the various components involved during the methanol synthesis process in an IL are measured. These include the so-called syngas components (CO, N_2, CO_2, and H_2) and the condensable methanol and water vapors in order to determine their ideal and real separation selectivities under different temperature and high-pressure conditions relevant to the MeS process. Second, the efficiency of two different sweep liquids have been compared in the MR experiments: a high boiling point (B.P.) petroleum-derived solvent, tetraethylene glycol dimethyl ether (TGDE) and an ionic liquid, 1-Ethyl-3-methylimidazolium tetrafluoroborate ([Emim][BF4]). The choice of the TGDE and the IL as sweep liquids is because MeOH has high solubility in both of them, while the permanent gases like H_2 and CO do not. Compared to TGDE, the IL has higher decomposition temperature (400 ^{o}C), which implies less demanding downstream processing requirements for the separation of the MeOH from the solvent; its lower vapor pressure means lower solvent loss, and its higher thermal capacity affords a broader range of MeS operating conditions. The TGDE, on the other hand, is at present more readily available and less costly In this study, the conversion and yield of the MeS reaction in a MR using IL as the sweep liquid was measured at different conditions and the results were compared with the MR system employing TGDE as the sweep liquid.

The experimental set-up consists of the syngas delivery system, the MR, the liquid injection system employing a high-pressure HPLC pump, and the gas and liquid analysis sections utilizing GC/TCD and GC/FID instruments. A commercial mesoporous alumina membranes, whose surface is rendered hydrophobic via the application of an appropriate modifying agent, is installed in the reactor. During operation, the syngas mixture is fed into the membrane shell-side that operates under high pressure and temperature, where it contacts the MeS Cu-ZnO-Al_2O_3 catalyst to convert into MeOH. The sweep liquid is pumped through the membrane tube-side via a HPLC pump to continuously remove in situ the produced MeOH, and thus help increase the conversion and yield of the MeS reaction. Experiments were conducted in a broad range of pressures and temperatures, sweep liquid flow rates and for two different sweep liquids, TGDE and IL. The lab-scale system operates also as a conventional packed-bed reactor (PBR) when the inlet and outlet of the membrane tube-side are closed, so as to compare its behavior to that of the MR.

The experiments in the project, which will be reported at a greater detail at the meeting, investigate the performance of the MR under different pressures, sweep liquid flow rates and syngas feed compositions (expressed in terms of the stoichiometric number SN = $(F_{H2}-F_{CO2})/(F_{CO}+F_{CO2})$ and the carbon factor CF = $F_{CO}/(F_{CO}+F_{CO2})$) and compare the carbon conversion of the MeS reaction in the MR with that of the PBR under the same conditions. For the whole range of pressures and flow rate studied, the IL attains higher conversions than the TGDE, which can be related to its higher methanol solubility. This result is in good agreement with the results of solubility tests which have revealed that methanol has a higher solubility in the IL compared to TGDE.

Constrains with the size of the lab-scale system limit the gains in conversion afforded by the MR system. Companion modeling studies, however, indicate that conversions >90% are readily attainable with the proposed process. Further, economic evaluations indicate that such conversions would allow the proposed process to operate on biomass-derived syngas in an one-pass configuration without recycle of unreacted syngas needed.

571412 Sulfuric Acid-Catalyzed Isobutane Alkylation with C4 Olefin Promoted By Sulfonic-Acid-Functionalized Ionic Liquids

Sulfuric Acid-Catalyzed Isobutane Alkylation with C4 Olefin Promoted by Sulfonic-Acid-Functionalized Ionic Liquids

Weizhen Sun, Piao Cao, and Ling Zhao

State Key Laboratory of Chemical Engineering, East China University of Science and Technology, Shanghai 200237, China

The alkylate produced by the C4 alkylation reaction with a high-octane number, no content of aromatics and sulfur and clean burning, is an ideal blending component for gas upgrading. The current worldwide alkylation mainly relies on the concentrated H_2SO_4 as the main commercial catalyst, which has high catalytic activity and selectivity while suffers from equipment corrosion, and environmental pollution. To overcome these problems in the H_2SO_4-catalyzed alkylation, the green and alternative catalysts or co-catalysts need to be developed for the C4 alkylation process.

As a dual solvent-catalyst, sulfonic-acid-functionalized ionic liquids (SFILs) are promising catalysts for the C4 alkylation because of their considerable advantages, including the high-catalytic activity and stability, easy separation from the product, strong Brønsted acidity, negligible-vapor pressure and designable structures. However, few work related to the effect of the SFILs on the interfacial behaviors of the H_2SO_4-catalyzed alkylation has been focused on. In addition, for the SFIL/H_2SO_4-catalyzed alkylation, the addition ratio of the SFIL to H_2SO_4 is the first important factor that needs to be studied, because the addition amount of the SFIL determines the interfacial properties as well as the acidity of the C4 alkylation process, and further determines the quality of alkylate. In this work, we focus on the effect of the 1-octyl-3-sulfopropyl-imidazolium hydrogen sulfate ([OPSim][HSO_4]) with different ratios to the H_2SO_4 on the catalytic performance and interfacial properties for the H_2SO_4-catalyzed alkylation by the combination of experiments and MD simulation.

Experimental results indicated that compared to pure H_2SO_4, the introduction of the SFILs can contribute to a higher selectivity of C8-alkylates as well as Research Octane Number (RON), suggesting a higher quality of alkylate. Moreover, the catalytic performance of the SFIL/H_2SO_4 catalyst is improved with the increased addition ratio of the SFIL into the H_2SO_4 within the range of 10 wt%. MD simulation results reveal that the obvious density enrichment of the cations of the SFILs at the interface results in a larger interface width, and higher interface number and better dispersion of C4 hydrocarbons at the interface, which can be ascribed to the location of the alkyl chains close to the C4 hydrocarbon phase with the extended and perpendicular behaviors. The improvement of interfacial properties due to the addition of the SFILs is favorable to the enhancement of the quality of alkylate, which is in excellent agreement with experimental results. Hopefully, the useful information in this work can provide a reference to the design of the SFILs and the optimization of the C4 alkylation process.

571207 Composite Adsorbents with Anion-Pillared Microporous Materials Embedded in Poly (Ionic Liquids) for Flue Gas Desulfurization

As one of the most notorious air pollutants, sulfur dioxide (SO_2) has been corroding the natural ecology and the human society since the onset of fossil fuel combustion. It is also one of the initiators for the formation of inhalable particle matters that are hazardous to human health. Hence, technologies to reduce the SO_2 emission are under intensive research.

Flue gas is one of the main SO_2 sources. Though fuel desulfurization technologies have been under rapid development, elimination of SO_2 from flue gas is still necessary due to the enormous amount of emission. Traditional processes involving alkaline solution or limestone will produce extra wastes. Adsorption, however, can be waste-free and even provide the possibility of SO_2 recovery. Thus, new materials with impressive adsorption performances are being designed. Since flue gas usually contains only about 0.2%(v/v) of SO_2, the adsorbents are required to grab SO_2 at very low partial pressures. Meanwhile, with around 10%(v/v) CO_2 in the gas mixture, the adsorbents should also possess high SO_2/CO_2 selectivity to avoid frequent regeneration. Metal-organic frameworks (MOFs) have shown sufficiently high SO_2 adsorption capacity. However, these materials also take up CO_2 readily, resulting in limited SO_2/CO_2 selectivity.

Herein, we present a novel composite material consisting of anion-pillared microporous materials and poly (ionic liquids) for flue gas desulfurization. As an attractive class of MOFs, anion-pillared microporous materials exhibited extremely high SO_2 adsorption capacities at low partial pressures, as revealed by our previous work. Poly (ionic liquids), on the other hand, showed distinctively high SO_2/CO_2 selectivity via particular swelling toward SO_2. By embedding anion-pillared materials into poly (ionic liquids), we obtained a new class of composite adsorbents able to grab SO_2 at an extremely low partial pressure while selectively repulse CO_2, which is especially adequate for flue gas desulfurization. The strategy described here can also be applied to other adsorbents for separation performance improvement.

Acknowledgement:

This research was supported by the National Natural Science Foundation of China (No. 21725603), Zhejiang Provincial Natural Science Foundation of China (No. LZ18B060001), and the National Program for support of Top-notch Yong Professionals (H. X.)

References:

1. X. L. Cui, Q. W. Yang, L. F. Yang, R. Krishna, Z. G. Zhang, Z. B. Bao, H. Wu, Q. L. Ren, W. Zhou, B. L. Chen, H. B. Xing. Adv. Mater. 2017, 29, 1606929.

2. L. Xia, Q. Cui, X. Suo, Y. H. Li, X. L. Cui, Q. W. Yang, J. H. Xu, Y. W. Yang and H. B. Xing. Adv. Funct. Mater. 2018, 28, 1704292.

567659 The Catalytic Mechanisms and Performances of Ionic Liquids in Green Reaction Processes

The catalytic mechanisms and performances of ionic liquids in green reaction processes

Xiaoqian Yao, Suojiang Zhang*

CAS Key Laboratory of Green Process and Engineering, Beijing Key Laboratory of Ionic Liquids Clean Process, State Key Laboratory of Multiphase Complex Systems, Institute of Process Engineering, Chinese Academy of Sciences, Beijing 100190, PR China

Email: sjzhang@ipe.ac.cn; xqyao@ipe.ac.cn

Ionic liquids are with many attractive properties including excellent catalytic and dissolvability properties, and in past few decades their huge potential are realized to replace the traditional organic solvent in many kinds of reactions. [1,2] Ionic liquids were supposed to be reaction intermediate and catalyst in different reaction processes, such as hydrogenation and carbonylation, etc. [3] Ionic liquid catalysts were developed and investigated extensively. However, absence of in-depth investigation on the reaction mechanism and performance of ionic liquid catalysis is a bottleneck problem in applications.

Based on the fundamental understanding in the structures and properties of ILs, several green reactions using ILs as catalysts or solvents, such as cleaner process of alkylation, formation of carbonates from CO_2, biomass conversion and PET polyester were investigated by quantum chemical calculations and multiscale simulations. The structure-activity relationship of ions and ionic clusters in ionic liquid catalysts were tried to be established. The synergistic law of cation and anion of was obtained, such as in reaction of CO_2 with epoxide and degradation of polyester catalyzed by ionic liquids. The catalysis mechanisms by ionic clusters in producing carbonate and branched alkanes are revealed, too. All of these researches would provide scientific basis for understanding the ionic liquid reactions, development of efficient ionic liquid catalyst, and establishment of green reaction processes.

Acknowledgments

This work was supported by National Natural Science Foundation of China (21776281), Natural Science Foundation of Beijing, China (2182073). and the National Key R & D Program of China (2017YFB0307303).

[1] Hallett, J. P.; Welton, T., Room-Temperature Ionic Liquids: Solvents for Synthesis and Catalysis. 2. *Chem. Reviews* 2011, 111, 3508-3576.

[2] Mahmoud, E.; Yu, J.; Gorte, R. J.; Lobo, R. F., Diels-Alder and Dehydration Reactions of Biomass-Derived Furan and Acrylic Acid for the Synthesis of Benzoic Acid. *ACS Catalysis* 2015, 5, 6946-6955.

[3] Bobbink, F. D.; Vasilyev, D.; Hulla, M.; Chamam, S.; Menoud, F.;Laurenczy, G.; Katsyuba, S.; Dyson, P. J., Intricacies of Cation-Anion Combinations in Imidazolium Salt-Catalyzed Cycloaddition of CO_2 Into Epoxides. *ACS Catalysis* 2018, 8, 2589-2594.

AUTHOR INDEX

AUTHOR INDEX

AUTHOR INDEX

AIChE
120 Wall Street, FL 23
New York, NY 10005-4020

ISBN 978-1-7138-0564-9

CURRAN ASSOCIATES INC.
proceedings.com